Programming with STM32

Getting Started with the Nucleo Board and C/C++

Programming with STM32

Getting Started with the Nucleo Board and C/C++

Donald Norris

New York   Chicago   San Francisco
Athens   London   Madrid
Mexico City   Milan   New Delhi
Singapore   Sydney   Toronto

Library of Congress Control Number: 2017959742

Programming with STM32 Getting Started with the Nucleo Board and C/C++

123456789   LCR   22 21 20 19 18

ISBN   978-1-260-03131-7
MHID   1-260-03131-4

The pages within this book were printed on acid-free paper.

Sponsoring Editor	**Project Manager**	**Indexer**
Michael McCabe	Nikhil, MPS Limited	Edwin Durbin
Editorial Supervisor	**Copy Editor**	**Art Director, Cover**
Donna M. Martone	Md. Taiyab Khan, MPS Limited	Jeff Weeks
Production Supervisor		**Composition**
Lynn M. Messina	**Proofreader**	MPS Limited
Acquisitions Coordinator	A. Nayyer Shamsi, MPS Limited	
Elizabeth Houde		

About the Author

Donald J. Norris has a degree in electrical engineering and an MBA specializing in production management. He is currently an adjunct professor teaching an Embedded Systems course in the College of Engineering, Technology and Aeronautics, part of the Southern New Hampshire University (SNHU). He has also taught many different undergrad and grad courses mainly in the computer science and technology areas at SNHU and other regional schools for the past 33 years. Don created and taught the initial robotics courses at SNHU both on-campus and online.

Don retired from civilian government service with the U.S. Navy, where he specialized in underwater acoustics related to nuclear submarines and associated advanced digital signal processing systems. Since then, he has spent more than 23 years as a professional software developer using the C, C#, C++, Python, Micro Python, Node.JS, JavaScript, PHP, and Java languages in varied development projects. He also has been a certified IT security consultant for the last six years.

He has written and had published seven books including three involving the Raspberry Pi, one on how to build and fly your own drone, a book on the Intel Edison, one on the Internet of Things, and one on Micro Python.

Don started a consultancy, Norris Embedded Software Solutions (dba NESS LLC), which specializes in developing application solutions using microcontrollers, sensors, and actuators. The business has also recently completed several robotics projects for clients.

Don likes to think of himself as a perpetual hobbyist and geek and is constantly trying out new technologies and out-of-box experiments. He is a licensed private pilot, photography buff, amateur extra class operator, avid runner, and most importantly, a proud grandfather of three great kids, Evangeline, Hudson, and Holton.

This book is dedicated to Dr. Peter Kachavos, my son-in-law, who is a remarkably intelligent man with an equally remarkable long medical career in service to his patients and the community. Until recently, Peter was a practicing internist with an office in Manchester, NH. He recently retired after 25 years from that practice and soon will be pursuing other interesting opportunities in the medical field.

Peter enjoys cooking, fine wine, traveling, and spending quality time with his family. His wife is my daughter, Shauna, and their child is my two-year-old granddaughter, Evangeline.

Peter and I have spent many hours discussing many topics ranging from ancient Greek artifacts to the latest technologies impacting modern society. I always look forward to those interesting and challenging discussions.

CONTENTS AT A GLANCE

1   Introduction to the STMicroelectronics
Line of Microcontrollers   1

2   STM MCU Software   29

3   STM32CubeMX Application   53

4   STM Project Development   87

5   General-Purpose Input Output (GPIO)
and the STM Hardware Abstraction Layer (HAL)   101

6   Interrupts   129

7   Timers   145

8   Bit Serial Communications   167

9   Analog-to-Digital Conversion   189

10   Pulse Width Modulation (PWM)   213

11   Direct Memory Access (DMA) and
the Digital-to-Analog Converter (DAC)   241

Index   281

CONTENTS

Preface xiii

1 Introduction to the STMicroelectronics
Line of Microcontrollers 1
Microcomputer vs Microcontroller 1
STM Nucleo Boards 2
Principal MCU Components 3
Bit Serial Ports 19
Nucleo-64 Board Options 22
Summary 27

2 STM MCU Software 29
Open-Source versus Commercial Proprietary Software 29
Bare Metal Development 30
Brief History of MCU 30
The MCU Toolchain 32
Configuring a STM32 Toolchain 34
Summary 52

3 STM32CubeMX Application 53
Pinout Tab 58
MCU Alternative Functions 60
Integrated Peripheral (IP) Tree Pane 61
Creating an Example Project using CubeMX 66
The main.c Code Listing 72
ARM Cortex Microcontroller Software Interface
Standard (CMSIS) 78
CubeMX-Generated C Code 80
Compiling and Downloading the Project 80
Downloading the Hex Code 82
Summary 86

4 STM Project Development 87

Hello World Project 87

Creating the Hello Nucleo Project 87

Adding Functionality to the Program 98

Compiling and Executing the Modified Program 98

Simple Modification for the main.c Function 99

Complex Modification for the main.c File 99

Summary 100

5 General-Purpose Input Output (GPIO)
and the STM Hardware Abstraction Layer (HAL) 101

Memory-Mapped Peripherals 103

Core Memory Addresses 103

Peripheral Memory Addresses 105

HAL_GPIO Module 107

GPIO Pin Hardware 109

LED Test Demonstration 111

Enabling Multiple Outputs 122

Push-Button Test Demonstration 123

Clock Speed Demonstration 124

Setting the Pin Clock Speeds 125

Summary 128

6 Interrupts 129

Interrupts 129

NVIC Specifications 130

Interrupt Process 131

External Interrupts 134

Interrupt Demonstration 135

Summary 144

7 Timers 145

STM Timer Peripherals 145

STM Timer Configuration 147

Update Event Calculation 148

Polled or Non-interrupt Blink LED Timer Demonstration 149

Test Run 155

Interrupt-Driven Blink LED Timer Demonstration 155

Test Run 157

Multi-rate Interrupt-Driven Blink
LED Timer Demonstration 157

Test Run 164

Modification to the Multi-rate Program 165

Test Run 165

Summary 166

8 Bit Serial Communications 167

UARTs and USARTs 167

USART Configuration 170

Windows Terminal Program 172

Enabling USART2 175

USART Transmit Demonstration Program 176

Test Run 181

USART Receive Demonstration Program 182

Test Run 187

Summary 188

9 Analog-to-Digital Conversion 189

ADC Functions 190

ADC Module with HAL 192

ADC Conversion Modes 197

Channels, Groups, and Ranks 198

ADC Demonstration 200

ADC Demonstration Software 205

Summary 212

10 Pulse Width Modulation (PWM) 213

General-Purpose Timer PWM Signal Generation 214

Timer Hardware Architecture 216

PWM Signals with HAL 216

Enabling the PWM Function 220

PWM Demonstration Software 222

Demonstration One 227

Demonstration Two 228

Demonstration Three 231

Demonstration Four 235

Adding Functional Test Code 239

Test Results 239

Summary 240

11   Direct Memory Access (DMA) and
the Digital-to-Analog Converter (DAC) 241

DMA 241

Basic Data Transfer Concepts 242

DMA Controller Details 245

Using HAL with DMA 246

Demonstration One 251

DAC Peripheral 259

DAC Principles 259

HAL Software for the DAC 261

Demonstration Two 263

Demonstration Three 269

Summary 278

Index 281

PREFACE

This book will serve both as an introduction to the STMicroelectronics line of STM32 microcontrollers (MCUs) and also as an easy-to-follow Getting Started Guide for readers interested in developing with a STM MCU. I will be using one of the very inexpensive STM Nucleo-64 development boards for all of the book projects, which should make it inviting for most readers to become involved with the hardware. In fact, doing the book demonstration projects is really the only way you can really be assured that you have gained a good comprehension of the material in this book.

I will state from the beginning that it is simply not possible to gain a total understanding of how a STM MCU functions by only reading this book. The manufacturer datasheets that describe individual STM MCUs are often over 1,000 pages in length, which describes the enormity of the task of trying to master the voluminous amount of information that describes these devices. Instead, the book contents focus on a few of the core components that make up a STM MCU and how to program those components to accomplish fairly simple tasks.

Some readers will have trepidation about starting to develop with what are typically considered professional grade MCUs. I wish to allay that fear and state that I have found that developing applications with at least one representative sample STM MCU to be remarkably easy and straightforward. In fact, I will state that in some aspects it is easier to develop with a STM MCU than with an Arduino or Raspberry Pi, which many readers will already be quite familiar and probably have already created projects with those boards.

Often, the single biggest issue with developing with MCUs is setting up a stable development toolchain. I will describe how to do this in a simple to follow, step-by-step process, which if you rigorously follow will guarantee that you will be able to quickly and without much trouble generate working binary programs. These programs will then be quickly downloaded into the development board for execution.

1

Introduction
to the STMicroelectronics
Line of Microcontrollers

This chapter provides you with an introduction to the very comprehensive STMicroelectronics (STM) line of microcontrollers (MCUs). I will be focusing only on several specific controllers throughout the book, but that should provide you with an adequate representation of the functions and capabilities of the full line of STM MCUs.

Microcomputer vs Microcontroller

I believe at the start of this book that it must make very clear the differences between a microcomputer and a MCU. The reason for this distinction is very simple: STM is a company that designs and manufacturers MCUs, not microcomputers. I think my following definition of a MCU is as good as any that I have read:

> *A microcontroller is an integrated system containing a minimum of a microprocessor, dynamic and non-volatile memory, and a set of peripherals consistent with all design requirements.*

Right away, you can see from the definition that a MCU contains a microprocessor which is sometimes referred to as a microcomputer. There also must be both dynamic or volatile memory as well as nonvolatile or static memory, where

1

the latter holds any programs or scripts necessary to run the microprocessor. Finally, there are always peripherals added to the design that allow for the input and output of digital signals. There are often additional peripherals such as timers, interrupt controllers, serial data ports, and a variety of others depending upon what requirements the MCU must meet.

I discuss all the principal MCU components below to provide you with a solid background to understand how a MCU functions. Most of the following explanations are based on the voluminous amount of information provided by STM on their microcontrollers. The reason behind STM providing such a large amount of information is to allow engineers/software developers access to all the data they need in order to incorporate STM products into original equipment manufacturer (OEM) designs and products. This approach is fundamentally different than the approach taken by suppliers of maker style boards and products, such as the Arduino, Raspberry Pi, BeagleBone, and so forth. In the case of the latter, board documentation is geared toward how to use a board in a project. The STM data is extremely specific describing items such as the nanosecond timing pulses between memory chips and processor buses. This is exactly the reason why some of the STM MCU datasheets are over 1,100 pages in length. Fortunately, the datasheet for the primary STM MCU I will be using in this book is only 138 pages. I will provide later in this chapter the website where you can download the datasheet.

STM Nucleo Boards

MCU manufacturers such as STM have long recognized that they just couldn't provide only chips to potential customers as most would have no way to effectively evaluate them for potential use in their products. This is the reason why the manufacturers offer relatively low-cost evaluation and demonstration boards, which have representative MCUs all setup and ready to run. STM offers a series of such boards that it calls the Nucleo line. I will be using a fairly simple Nucleo-64 board for the book projects. STM has actually embraced the maker community by marketing the Nucleo lineup as boards suitable for maker project use. These boards are very inexpensive, usually about US$10–15, which leads me to believe that STM is striving to gain a foothold in the maker community by actually subsidizing the manufacturing costs for the boards. In any case, this is a boon for makers and hobbyists and one that we should embrace.

Figure 1-1 shows the three basic Nucleo boards available at the time of writing this book.

Figure 1-1   *Three basic Nucleo boards.*

The boards are named Nucleo-32, Nucleo-64, and Nucleo-144 from left to right, respectively, as shown in Figure 1-1. The number in each name represents the number of pins present in the MCU chip. Nucleo-64 is the principal board used in this book.

Principal MCU Components

The first component to consider is the processor or microcomputer.

Processor

The processor used in Nucleo-32 and Nucleo-64 boards is the ARM Cortex M-4 32-bit processor. It too has a lengthy 278-page user guide available from infocenter. arm.com/help/topic/com.arm.doc.dui0553a/DUI0553A_cortex_m4_dgug.pdf. The actual processor circuitry is part of the STM MCU because STM has purchased intellectual property (IP) rights from the ARM Corporation in order to integrate it into its chips. However, for all practical purposes, the ARM processor is programmed using the tools and techniques promulgated by the ARM Corporation to support its processor IP. This distinction is of no consequence in our case because the software tools to be used for the book projects have all been carefully crafted and tested to work seamlessly together by STM. Any license issues have already been resolved without bothering the end user.

The following list contains some of the important specifications for the Cortex M-4 processor for interested readers:

- Full-featured ARMv7-M instruction set, optimized for embedded applications

- Floating point unit (FPU)

- Low-power 32-bit processor

- Memory protection unit (MPU)

- Nested vector interrupt controller (NVIC)

- Trace, breakpoint, and JTAG capabilities

- Advanced Microcontroller Bus Architecture (AMBA)

- Advanced High-Performance Bus (AHB5, *AHB-Lite*)

There are many more features to the Cortex M-4 processor as the 278-page user guide would suggest. I would also recommend Joseph Yiu's book, *The Definitive Guide to ARM® Cortex®-M3 and Cortex®-M4 Processors*, Third Edition, to readers who really want to delve into this processor to a great depth.

Figure 1-2 is a block diagram of the Cortex M-4 processor showing all of the listed items and more.

STM is responsible for the design and implementation of all components outside of the box shown in Figure 1-2. Of course, the STM design must be compliant with the ARM processor specifications. One item missing in the figure is a clock input. The reason for it being missing is that the block diagram is from the ARM Cortex M-4 user guide, while the clock circuitry is the responsibility of STM, the MCU designer. STM has set the base processor clock frequency at 72 MHz, which appears to be very low in today's world of multi-gigahertz clock rates for most PCs. In fact, the latest Raspberry Pi model 3 has a 1.2-GHz clock rate. But first appearances are deceiving in this case. The Cortex M-4 processor uses an extremely efficient instruction set, with three-stage pipelining, which maximizes the performance of the underlying reduced instruction set computing (RISC) that the processor employs. In addition, many common microcontroller tasks, which I discuss below, have been implemented in a combination of hardware and

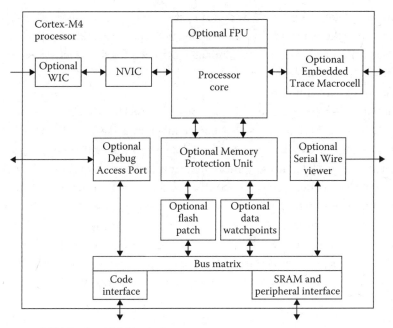

Figure 1-2 *ARM Cortex M-4 block diagram.*

firmware, further improving the overall performance, while keeping the power consumption as low as possible. Power consumption and performance are two key microcontroller attributes that system designers always keep in mind. Keeping the clock rate as low as possible will always minimize the power consumed as well as increase the longevity of the chip.

Memory

The Cortex M-4 processor uses a Harvard architecture. This means that program instructions are stored and retrieved from a memory separate and distinct from the memory that holds data. The other common computer architecture is named von Neumann in which instructions and data share a common memory. MCUs have very limited memory and take advantage of the speed-up available by using concurrent instruction and data access. Additionally, having separate memories means that the processor is no longer constrained to the same sized data widths. This means instructions can be fetched and executed in 4-byte chunks or 32 bits, while data can be simultaneously handled with 1-byte or 8-bit chunks, thus speeding up the overall throughput to and from the processor and memory. Another advantage the Harvard architecture has over the von Neumann form is that instruction prefetches can now be done in parallel with regular instruction executions, thus further speeding up the overall system performance. Finally, concurrent instruction and data access eliminates the need for data caches, which are typically used in von Neumann machines. This further reduces system complexity and power consumption.

The Nucleo-64 board used in this book is the STM MCU, model number STM32F302R8. This chip has a 64-KB flash memory and a 16-KB static random access memory (SRAM). Yes, those are kilobytes, not mega- or gigabytes. You cannot expect to create any graphical programs that run in this limited memory space. Microcontroller programs are truly a throwback to the earliest days of computing where memory was very limited and developers had to use every available byte to store and execute programs. Having mentioned the scarce memory resources, you will be able to use a variety of modern-day graphic-based programs to develop the MCU program, but they will run on a PC. The compiled and optimized binary code will be downloaded into the MCU from the PC.

It turns out that 64 KB is plenty of space to run fairly large programs because the C/C++ cross-compiler used for this book's projects produces optimize code, eliminating all but the essential instructions needed for the program. The 16-KB SRAM size is more than adequate for the dynamic memory requirements to support a maximum 64-KB sized program. The actual memory is integrated onto the MCU chip and its type is largely irrelevant for our purposes.

Peripherals

The peripherals are what make the MCU viable for its intended purposes. In my microcontroller definition I stated, "a set of peripherals consistent with all design requirements." This means that microcontrollers typically have different peripheral configurations depending upon the requirements they are designed to meet. This is a major reason why STM manufactures such a wide variety of MCUs. I will focus on the Nucleo-64 microcontroller I mentioned above and will use its set of peripherals as my discussion points. I will next need to demonstrate how the Nucleo-64 board is set up in order to fully explain how all the peripheral features function with this board.

Nucleo-64 Board Layout

Figure 1-3 is a block diagram showing how the major components that make up a Nucleo-64 board are configured.

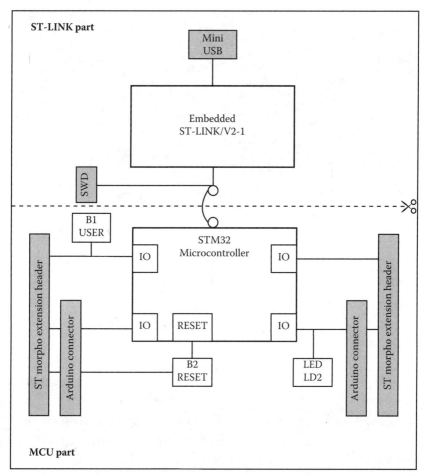

Figure 1-3 *Nucleo-64 block diagram.*

The first item you should note is that the board is actually made of two sections, the top section being a ST-LINK/V2-1 programmer and the bottom section is the STM32 MCU. The bottom section is also fitted with two sets of connectors. One set is compatible with the Arduino shield pin configuration and other set is the ST Morpho pinout configuration. There are two push buttons, B1 and B2, shown in the figure. B1 is multipurpose, while B2 is the reset push button. There is also one LED indicated in the figure; however, there are several others also which are shown in follow-on figures. The programmer section may be separated from the micro-controller section by snapping the sparse printed circuit board (PCB) joints, which can be seen in Figure 1-1. I would not recommend doing so because you would lose

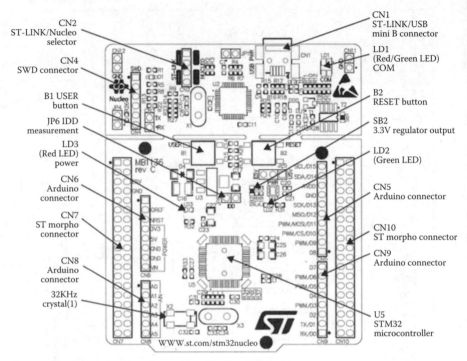

Figure 1-4   *Annotated top view of the Nucleo-64 board.*

all programming capability for the MCU section. I suppose it was set up in this fashion so you could separate the sections once there was a working MCU for a specific project and there was no need for any further programming.

Figure 1-4 is a top view of the Nucleo-64 board with most of the components clearly annotated.

Remarkably, there are relatively few components on the board. Most of the space is taken up by connectors, push buttons, configuration jumpers, LEDs, and the microcontroller chip. This sparse layout reflects the nature of the MCU, which is to be an embedded device without any of the niceties needed for the human-computer interface (HCI). HCI is possible with this board by using an appropriate extension board that plugs into this board

Figure 1-5 shows the bottom side of the Nucleo-64 board.

Not much to see in this figure other than a bunch of 0-ohm resistors that may be removed if it was desired to disable certain features such as the reset and user

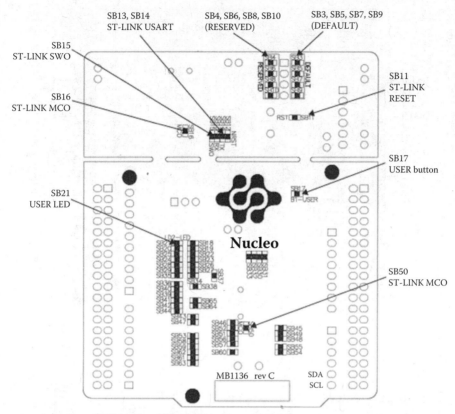

Figure 1-5 *Annotated bottom view of the Nucleo-64 board.*

buttons and user LED, or make hardware configuration changes to the ST-LINK programmer.

Figure 1-6 is much more interesting as it shows the pinouts for all the connectors on the top of the Nucleo-64 board.

There are two sets of pinouts shown in Figure 1-6, one belonging to the Arduino and the other to the ST Morpho. Connectors CN5, CN6, CN8, and CN9 belong to the Arduino set. Connectors CN7 and CN10 belong to the Morpho set. It is important to note that there are no one-to-one relationships between the sets. The Morpho connectors are two double rows of 38 pins for a total of 76 pins. The Arduino connectors are single-row connectors totaling 32 pins in all spread across four connectors.

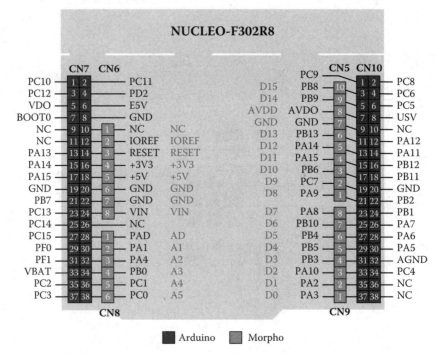

Figure 1-6   *Nucleo-64 pinout diagram.*

Arduino Connectors

Tables 1-1 to 1-4, which are from the Nucleo-64 User Guide, detail the Arduino connectors with cross-references between Arduino pin terms and STM32 pins. I have provided further clarifications on some of the pin functions following each table, where I felt it was appropriate.

Connector	Pin	Pin Name	STM32 Pin	Function
		Left connector		
CN6 power	1	NC	–	–
	2	IOREF	–	3.3V
	3	RESET	NRST	RESET
	4	3.3V	–	3.3V
	5	5V	–	5V
	6	GND	–	Ground
	7	GND	–	Ground
	8	VIN	–	Power input

Table 1-1   *CN6 Pinout*

Table 1-1 details the CN6 connector pinout. Refer to Figure 1-4 to see the placement of CN6 on the board.

Grounding pin 3 will reset the MCU. The STM32 pin designation NRST is short for non-maskable reset, which means a low on this pin is always recognized as a forcible MCU reset. Pins 2 and 4 provide 3.3 V, but I would recommend that only a very limited amount of current be drawn from these pins or you risk disrupting the board operations. Pin 5 provides 5 V, but again my same caution exists for excessive current draw. VIN is where you can attach an external power supply. The input voltage range is approximately 7 to 12 V based on the voltage regulator ratings connected to this pin. Jumper JP5 must be switched to accommodate the external power supply. Finally, the IOREF pin allows Arduino shields to adjust the I/O voltage to the 3V3 for the STM32 if those shields follow the STM specification.

Table 1-2 details the CN8 connector pinout. Refer to Figure 1-4 to see the placement of CN8 on the board.

These pins are the analog voltage inputs that are routed to the MCU multichannel analog-to-digital converter (ADC). You must restrict the analog input voltage to 3.3 V or less to avoid damaging the ADC inputs. Note that there is no one-to-one correspondence between the Arduino channels and the MCU channels. Pins 5 and 6 are the default ADC inputs. If you need these to bus #1 I2C inputs, you must disconnect solder bridges SB51 and SB56 and connect SB42 and SB46. These connections are made using 0-ohm resistors on the bottom side of the MCU board. Please note that the bus #1 I2C connections are already available on pins 1 and 2 on connector CN5.

Table 1-3 details the CN5 connector pinout. Refer to Figure 1-4 to see the placement of CN5 on the board.

Connector	Pin	Pin Name	STM32 Pin	Function
		Left connector		
CN8 analog	1	A0	PA0	ADC1_IN1
	2	A1	PA1	ADC1_IN2
	3	A2	PA4	ADC1_IN5
	4	A3	PB0	ADC!_IN11
	5	A4	PC1 or PB9	ADC1_IN7 or ISC1_SDA (PB9)
	6	A5	PB0 or PB8	ADC1_IN6 or I2C1_SCL (PB8)

Table 1-2 *CN8 Pinout*

Connector	Pin	Pin Name	STM32 Pin	Function
		Right connector		
CN5 digital	10	D15	PB8	I2C1_SCL
	9	D14	PB9	I2C1_SDA
	8	AREF	–	AVDD
	7	GND	–	Ground
	6	D13	PB13	SPI2_SCK
	5	D12	PB14	SPI2_MISO
	4	D11	PB15	TIM15_CH2 or SPI2_MOSI
	3	D10	PB6	TIM16_CH1N or SPI2_CS
	2	D9	PC7	–
	1	D8	PA9	–

Table 1-3 *CN5 Pinout*

The analog voltage reference (AREF) should be connected to 3.3 V because that is the maximum allowed voltage to be input to the MCU ADC channels. If you have another supply available with a lesser voltage, say 2.5 V, then you can connect AREF to that and increase the ADC dynamic range by a third. Just be aware that inputting a voltage greater than 2.5 V will not change the ADC output, effectively saturating it. Pins 3 to 6 may be configured as general-purpose input/output (GPIO) or act as the bus #2 serial peripheral interface (SPI). Pins 2, 3, and 4 can also act as timer pins if neither the GPIO nor the SPI is required.

Table 1-4 details the CN9 connector pinout. Refer to Figure 1-4 to see the placement of CN9 on the board.

These eight pins are normally regular GPIO, but they may be reconfigured so that pins 4, 6, and 7 are timer pins. In addition, pins 1 and 2 can act as the #2 USART.

Connector	Pin	Pin Name	STM32 Pin	Function
		Right connector		
CN9 digital	8	D7	PA8	–
	7	D6	PB10	TIM2_CH3
	6	D5	PB4	TIM16_CH1
	5	D4	PB5	–
	4	D3	PB3	TIM2_CH2
	3	D2	PA10	–
	2	D1	PA2	USART2_TX
	1	D0	PA3	USART2_RX

Table 1-4 *CN9 Pinout*

Morpho Connectors

Table 1-5 is from the Nucleo-64 User Guide and details the STM Morpho connections with references to the STM32 pins. I have provided further clarifications on some of the pin functions following the table, where I felt it was appropriate. I did not repeat the comments I provided for Arduino connection as they are equally valid for the Morpho connections.

Table 1-5 details the CN7 and CN10 connector pinouts. Refer to Figure 1-4 to see the placement of CN7 and CN10 on the board.

U5V is 5-V power supplied from the ST-LINK/V2-1 USB connector. The BOOT0 input pin allows for changing how the MCU initially boots during a power-on event. I will discuss the boot options in a later chapter. VBAT input is often used as a battery backup for powering a real-time clock or a security key. It is typically tied to VCC if not used for a battery backup.

The next series of MCU components that I will discuss enable the MCU to perform its operations more efficiently and quickly than possible by relying on software constructs.

CN7 Odd Pins		CN7 Even Pins		CN10 Odd Pins		CN10 Even Pins	
Pin	Name	Name	Pin	Pin	Name	Name	Pin
1	PC10	PC11	2	1	PC9	PC8	2
3	PC12	PD2	4	3	PB8	PC6	4
5	VDD	E5V	6	5	PB9	PC5	6
7	BOOT0	GND	8	7	AVDD	U5V	8
9	–	–	10	9	GND	–	10
11	–	IOREF	12	11	PB13	PA12	12
13	PA13	RESET	14	13	PB14	PA11	14
15	PA14	3.3V	16	15	PB15	PB12	16
17	PA15	5V	18	17	PB6	PB11	18
19	GND	GND	20	19	PC7	GND	20
21	PB7	GND	22	21	PA9	PB2	22
23	PC13	VIN	24	23	PA8	PB1	24
25	PC14	–	26	25	PB10	PA7	26
27	PC15	PA0	28	27	PB4	PA6	28
29	PF0	PA1	30	29	PB5	PA5	30
31	PF1	PA4	32	31	PB3	AGND	32
33	VBAT	PB0	34	33	PA10	PC4	34
35	PC2	PC1 or PB9	36	35	PA2	–	36
37	PC3	PC0 or PB8	38	37	PA3	–	38

Table 1-5 *CN7 and CN10 Pinouts*

Nested Vectored Interrupt Controller (NVIC)

Interrupts are essential elements in the MCU structure. Without them, MCUs cannot efficiently function given the limited memory available and low clock speeds. MCU interrupts are implemented both in hardware and software. But first I need to define what an interrupt is and how it is used:

> *An interrupt is a disruption in the normal flow of a program initiated by an external or temporal event.*

Figure 1-7 is a flow diagram of a generic interrupt sequence. This diagram will make it easier to understand how an interrupt works.

ISR in Figure 1-7 stands for interrupt service routine, which is the program code that executes whatever needs to be accomplished due to the interrupt. The block containing the phrase "Call ISR" causes the current state or context of the MCU to be saved in SRAM. The current state typically consists of the contents of program counter (PC) and a preset number of registers that depends on the MCU model. The PC is a register, which contains the address of the next instruction to be executed when the interrupt was recognized. The actual ISR code may be located anywhere in memory because it is accessed by replacing the PC with the first address of the ISR. That address is contained in a list, which is known as the

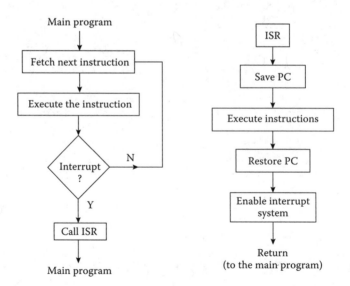

Figure 1-7 *Interrupt flow diagram.*

vector interrupt list. The last instruction in an ISR is always a return from interrupt, which restores the MCU's pre-interrupt state and resumes executing instructions at the stored PC address.

The figure is slightly off target and shows a block in the ISR section containing the phrase "Enable interrupt system." The STM NVIC uses prioritized interrupts, which means a higher priority interrupt can always interrupt a lower priority interrupt while it is executing. This normally is not possible with a purely software-implemented interrupt scheme, which is depicted by the figure's flow diagram.

Figure 1-8 shows two ways the NVIC handles multiple interrupt requests (IRQs).

I don't believe it makes much difference on how the NVIC handles interrupts for most non-safety-related projects. However, it would be very important if a designer was creating a safety critical system. One case that came to mind was an automotive control system, in which a STM MCU was managing a variety of systems including the infotainment system and a collision avoidance system. I certainly think that a collision avoidance IRQ would need immediate attention if the MCU was processing a radio station change request at the time the higher priority IRQ was received.

The NVIC used in the Nucleo-64 MCU project board has the following specifications:

- Sixteen interrupt lines

- Sixteen priority levels

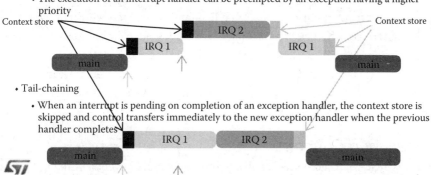

- Preemption and interrupt nesting
  - The execution of an interrupt handler can be preempted by an exception having a higher priority
- Tail-chaining
  - When an interrupt is pending on completion of an exception handler, the context store is skipped and control transfers immediately to the new exception handler when the previous handler completes

Figure 1-8 *NVIC handling multiple IRQs.*

- Up to 62 maskable interrupt channels
- Low-latency interrupt processing
- Processing of late arriving, higher priority interrupts
- Supporting tail chaining
- Interrupting entry restored on interrupt exit with no instruction overhead

I will discuss how to program ISRs in the software chapter.

General-Purpose Inputs/Outputs (GPIOs)

Each of the 51 GPIO pins can be configured by software as output with push-pull or open-drain and with or without pull-up or pull-down. They can also be configured as floating input, with or without pull-up or pull-down. Most GPIO pins have peripheral alternate functions as can be seen in the above tables. All GPIOs are high-current-capable and have limited speed selection to better manage internal noise, power consumption, and electromagnetic emission.

Analog-to-Digital Converter (ADC)

There is one 12-bit ADC that has 16 channels. ADC conversions can be done in either a single-shot or scan mode. In the scan mode, an automatic conversion is performed on a preselected group of analog inputs. An interrupt can be generated if a converted voltage is outside preset threshold. In addition, a specific channel conversion can be triggered by the TIM1, TIM2, TIM3, TIM4, or TIM5 timers.

External Interrupt/Event Controller (EXTI)

There is an external interrupt/event controller that uses 23 edge-detector lines to generate interrupt/event requests. Each line may be independently configured to select the trigger event such as leading edge, trailing edge, or both. All these lines can be independently masked. There is a register that maintains the status of each interrupt request. The EXTI can detect an external line with a pulse width shorter than the internal main clock period. Up to a maximum of 81 GPIOs can be connected to the 16 external interrupt lines.

Timers and Counters

Many typical MCU activities involve temporal or timing events. Using hardware timers and counters can minimize software execution cycles that are best allocated to other activities. That is why MCUs typically incorporate an array of

Timer Type	Timer	Counter Resolution	Counter Type	Pre-scalar Factor	Capture Compare Channels	Com-pare Output	Maximum Interface Clock (MHz)	Maximum Timer Clock (MHz)
Advanced control	TIM1	16-bit	Up, down, up/down	1 to 65,536	4	Yes	100	100
GP	TIM2	32-bit	up, down, up/down	1 to 65,536	4	No	50	100
GP	TIM15	16-bit	up	1 to 65,536	2	1	50	100
GP	TIM16,17	16-bit	up	1 to 65,536	1	1	100	100
Basic	TIM6	16-bit	up	1 to 65,536	0	No	100	100

Table 1-6 *MCU Timer Features*

timers and counters in their structure. Timers and counters are also tightly coupled with interrupts to provide an extremely efficient way to process timing and counting events.

Table 1-6 details the features supported by a single advanced control timer as well as the general purpose (GP) and basic timers.

Although there are two more watchdog timers, they are not listed in the table. I discuss each timer type below.

Advanced Control Timer (TIM1)

This timer has four independent channels that may be used for the following:

- Input capture
- Output capture
- Pulse width modulation (PWM)
- Single pulse output

It can be used as a GP 16-bit timer if desired; however, the PWM mode is especially useful for the control of standard and continuous rotation servos. This timer may also be linked with the GPx timers for event synchronization and chaining.

General-Purpose Timers (TIMx)

There are four GP timers available for use. These timers, TIM2, TIM15, TIM17, and TIM16, are full-featured timers with 16-bit prescalars. TIM2 is the only 32-bit timer, while the rest use 16 bits. All of these timers, except for the basic, have up to four channels for the following:

- Input capture
- Output compare
- PWM
- Single pulse output

These timers can work together in a linked manner and they can also link with TIM1.

Basic Timer (TIM6)

The remaining timer is TIM6, which is a basic 16-bit type. It has limited options and is used only for the most basic timing feature as its name implies.

Watchdog Timers

There are two watchdog timers. The first is an independent 12-bit, count-down with an 8-bit prescalar. It is driven by a separate 32-KHz crystal, which allows it to operate independently from the main MCU clock. It also operates when the MCU is in the standby or even stop mode. This allows it to reset the MCU if an event happens that inadvertently stops the MCU.

The second watchdog timer is a free-running 7-bit downcounter that can also be set up to reset the MCU in case a problem occurs. This watchdog runs from the MCU main clock, but it does have an early-warning interrupt capability, which gracefully stops the MCU in case of an unanticipated problem.

SysTick Timer

There is one more timer, SysTick, but it is not recommended to use. This timer is ordinarily dedicated to support real-time operating systems (RTOS). However, if the MCU project is not using an RTOS, then the timer is freely available. This timer features the following:

- A 24-bit downcounter
- Autoreload capability maskable system interrupt generation when the counter reaches 0
- Programmable clock source

I would say that this timer should only be used as a last resort in the unlikely case that all the other timers have been fully committed. I cannot conceive of a project complex enough where this would happen.

Bit Serial Ports

This MCU supports 4-bit serial protocols that allow for serial communication between it and other computing systems or peripherals.

USART Serial Protocol

The first bit-serial protocol to be discussed is the standard Universal Synchronous/Asynchronous Receiver/Transmitter (USART). This serial protocol may or may not require a clock signal depending upon if it is configured as synchronous or not. The MCU provides two USART terminal ports; however, only one named USART2 is active and it is directly connected to another USART port on a separate MCU located on the ST-LINK portion of the Nucleo-64 PCB. There is also pin pair CN3 located on the ST-LINK that exposes the TX and RX lines.

It is possible to disconnect the USART2 link by opening solder bridges SB13 and SB14 and closing solder bridges SB62 and SB63. If this is done, then the USART2 port on the main MCU using pins PA2 and PA3 would be connected to CN9, pins 1 and 2, as well as to CN10, pins 35 and 37, respectively. This would then provide a USART port to any compatible serial device located on an extension board.

A second USART port named USART3 can be enabled by using jumper wires between the ST Morpho connectors CN7 and CN3, located on the ST-LINK. For example, on this Nucleo-64 board it is possible to use USART3 available on PC10 (TX) and PC11 (RX). Two jumper wires would need to be connected as follows:

- PC10 (USART3_TX) available on CN7 pin 1 to CN3 pin RX

- PC11 (USART3_RX) available on CN7 pin 2 to CN3 pin TX

While this physically restores the USART link, I am fairly sure that the underlying ST-LINK communications software would have to be modified to use this new USART port number.

I2C Serial Protocol

The second bit-serial protocol is the Inter-Integrated Circuit interface or I2C (pronounced "eye-two-cee" or "eye-squared-cee"), which is also known as a

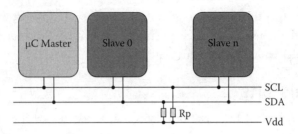

Figure 1-9 *I2C block diagram.*

Signal Name	Description	Connector/Pin Number
I2C1_SCL	Bus 1 Clock	CN5/10
I2C1_SDA	Bus 1 Data	CN5/9

Table 1-7 *I2C Signal Lines*

synchronous serial data link. A clock signal is needed because it is synchronous. Figure 1-9 is a block diagram of the I2C interface showing one master and one slave. This configuration is known as a multidrop or a bus network.

I2C supports more than one master as well as multiple slaves. This protocol was created by Philips in 1982 and is a very mature technology, meaning it is extremely reliable. Only two lines are used: SCLK for serial clock and SDA for serial data.

The STM MCU implements three separate I2C buses. These buses can operate at the standard clock rate of 100 KHz and the fast rate of 400 KHz. This clock rate can be increased by the manufacturer to a nonstandard 1 MHz. However, that would make the I2C implementation noncompliant with the protocol. Table 1-7 shows the Nucleo-64 pin designations for bus 1 clock and data lines.

SPI Serial Protocol

The third bit-serial protocol that I will discuss is the Serial Peripheral Interface (SPI), which is shown in Figure 1-10.

SPI (pronounced "spy" or "ess-pee-eye") is also known as a synchronous serial data link. It is also a full duplex protocol, meaning data can be simultaneously sent and received between the host and slave. SPI is also referred to as a Synchronous Serial Interface (SSI) or a 4-wire serial bus.

The MCU supports five SPI links in slave and master modes using full-duplex and/or simplex communication. The interfaces are labeled SPI1 to SPI5 and can

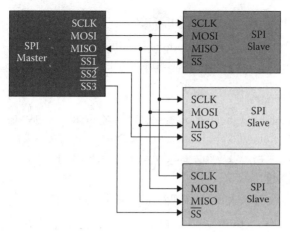

Figure 1-10 *SPI block diagram.*

Signal Name	Description	Connector/Pin Number and Name
MOSI	Bus 2 Master Out Slave In	CN5/6 PA7/D11
MISO	Bus 2 Master In Slave Out	CN5/5 PA6/D12
SCK	Bus 2 Clock	CN5/4 PA5/D13
/SS	Bus 2 Slave Select	CN5/3 PB6/D10

Table 1-8 *SPI2 Signal Lines*

use a clock rate of up to 25 Mbits/s. The data may be framed in either a byte or double-byte format. There is also automatic error correction in the form of CRC generation and verification.

Table 1-8 shows the Nucleo-64 pin designations for SPI2.

Inter-Integrated Sound

The fourth and final bit-serial interface to be discussed is the inter-integrated sound (I2S) interface. There are five standard I2S interfaces that are multiplexed with SPI1 to SPI5. I2S2 and I2S3 may be operated in a master or slave mode, in either simplex or full-duplex communication modes, and can be configured to operate with a 16/32-bit resolution as an input or output channel. All I2Sx interfaces can have audio sampling frequencies from 8 kHz up to 192 kHz. Finally, the master I2S clock can be output to the external DAC/CODEC at 256 times the sampling frequency when either I2S2 or I2S3 interface is configured in a master mode.

The above discussion is simply informational because I will not be demonstrating the I2S interface.

Nucleo-64 Board Options

The following sections discuss options specific to the Nucleo-64 board and are presented to provide you with some options when working with the board.

LEDs

LD1 is a tricolor LED located on the ST-LINK PCB and provides information concerning the state of communications between the ST-LINK, the host PC, and the main MCU. The following list shows the various LED states and the corresponding meaning:

- Slow blinking Red/Off: at power-on before USB initialization
- Fast blinking Red/Off: after the first correct communication between the PC and ST-LINK/V2-1 (enumeration)
- Red LED On: when the initialization between the PC and ST-LINK/V2-1 is complete
- Green LED On: after a successful target communication initialization
- Blinking Red/Green: during communication with target
- Green On: communication finished and successful
- Orange On: communication failure

LD2 is a green LED-labeled user and connected to Arduino signal D13 (CN5/CN6), which corresponds to STM32 I/O PB13

- When the I/O value is HIGH, the LED is on
- When the I/O value is LOW, the LED is off

LD3 is a red LED that indicates that the Nucleo-64 board is powered on and 5-V power is available.

Push Buttons

There are two push buttons on the board:

- B1 USER: The user button is connected to the I/O PC13 (CN7/23) of the MCU. This pin is pulled high through a 47K resistor. When B1 is pressed the pin is grounded.
- B2 RESET: This push button is connected to NRST, and is used to RESET the MCU.

NOTE   *The blue and black plastic hats that are placed on the push buttons can be removed if necessary, for example, when a shield or when an extension board is plugged on top of the Nucleo-64 board. This will avoid pressure on the buttons and consequently a possible permanent target RESET.*

JP6

Jumper JP6 is labeled IDD and is used to measure the MCU current by removing the jumper and connecting an ammeter in *series* with the two pin leads.

- Jumper ON: MCU is powered (default).

- Jumper OFF: An ammeter must be connected to measure the MCU current. The MCU cannot receive any power if there is no ammeter connected.

OSC Clock

There are four options on how to connect an high-speed external clock (HSE) on the Nucleo-64 board.

Master Clock Output from the ST-LINK (Default)

The master clock output (MCO) is generated by an 8-MHz crystal located on the ST-LINK PCB and input into the main MCU via the PF0/PD0/PH0-OSC_IN pins. This crystal frequency is fixed. Incidentally, the 72-MHz clock rate I mentioned earlier in this chapter is generated by using a phase-locked loop (PLL) embedded in the MCU using the 8-MHz input clock signal.

HSE Oscillator On-board with the Optional X3 Crystal

An 8-MHz crystal must be installed in empty X3 PCB pads. In addition, there are some solder bridges to be changed, and some resistors and capacitors to be added. The details are specified in the user's guide.

External Oscillator Through Pin 29 of the CN7 Connector

There are some solder bridges to be changed, and some resistors to be removed. The details are specified in the user's guide.

HSE Not Used

PF0/PD0/PH0 and PF1/PD1/PH1 are used as GPIOs instead of as a clock source. There are some solder bridges to be changed, and some resistors to be removed.

The details are specified in the user's guide. I am unsure how the MCU functions without an HSE, but it is an option according to the user's guide.

OSC 32-kHz Clock Supply

There are three ways to configure the pins corresponding to low-speed external clock (LSE).

LSE On-board Oscillator

You can change the X2 crystal provided that it conforms to the oscillator design guide for STM32 MCU. The default crystal already installed is a model ABS25-32.768KHZ-6-T manufactured by Abracon Corporation.

External Oscillator Through Pin 25
of the CN7 Connector

There are some solder bridges to be changed, and some resistors to be removed. The details are specified in the user's guide.

LSE Not Used

PC14 and PC15 are used as GPIOs instead of the LSE source. There are some solder bridges to be changed, and some resistors to be removed. The details are specified in the user's guide. I am fairly sure that the real-time clock and watchdog timers will become inoperative with the LSE disabled.

Power Supply and Power Selection

I have already discussed on where an external power supply may be attached in the Arduino connector. However, there are still some more power supply features and options you should know about.

Power Supply Input from the USB Connector

The whole Nucleo-64 board including any attached extension board can be powered by the 5-V power supplied through the CN1 USB connector provided the host USB device complies with the USB standard of providing 500 mA. The ST-LINK requires 100 mA and the main MCU requires 300 mA, which leaves 100 mA for any extension board or external devices such as LEDs or LCD displays. You must use an external supply if the total current is planned to exceed a nominal 500-mA load. The red LED LD3 will not light if there is insufficient current present to power the MCU.

When the board is power supplied by USB (U5V), a jumper must be connected between pins 1 and 2 of JP5.

Jumper JP1 is configured according to the maximum current consumption of the board when powered by the USB (U5V) connection. JP1 jumper can be set in case the board is powered by USB and maximum current consumption on U5V does not exceed 100 mA (including an eventual extension board or Arduino shield). In such condition, the USB enumeration will always succeed since no more than 100mA is required from the host. Possible configurations of JP1 are summarized in Table 1-9.

NOTE   *If the Nucleo-64 board is powered by an USB charger, there will be no USB enumeration from the ST-LINK causing the LD2 LED to remain off. In this case, JP1 needs to remain in place so that the MCU can be powered on.*

VIN and E5V External Power Supply Inputs

The details and limitations of external power sources VIN and E5V are summarized in Table 1-10.

When the board is power supplied by VIN or E5V, these jumper configurations must be set as follows:

• Jumper on JP5 pins 2 and 3

• Jumper removed on JP1

Jumper State	Power Supply	Allowed Current
JP1 jumper OFF	USB power through CN1	300 mA
JP1 jumper ON		100 mA

Table 1-9   *JP1 Configuration Table*

Input Power Name	Connector Pins	Voltage Range	Maximum Current	Limitations
VIN	CN6, pin 8; CN7, pin 24	7 to 12 V	800 mA	From 7 to 12 V only, and input current capability is linked to input voltage as follows: 800-mA input current when VIN = 7 V / 450-mA input current when 7 V < VIN ≤ 9 V / 250-mA input current or when 9V < VIN ≤12 V
E5V	CN7, pin 6	4.75 to 5.25 V	500 mA	–

Table 1-10   *VIN and E5V External Supply Input Limitations*

3.3-V external Power Supply Input

It is possible to power the MCU with an external 3.3-V power supply. The power connection would be made at CN6, pin 4 or CN7, pins 12 and 16. In this situation the 3.3-V supply is coming from an extension board. The ST-LINK will not be powered on with this configuration, thus making programming and debugging unavailable. The 3.3-V external power supply configuration is summarized by Table 1-11.

External Power Supply Output

When powered by USB, VIN, or E5V, the 5-V supply (CN6, pin 5 or CN7, pin 18) may be used as output power supply for an Arduino shield or an extension board. In this case, the maximum current of the power source specified in Table 1-10 must be respected. The 3.3 V (CN6, pin 4 or CN7, pins 12 and 16) can be used also as power supply output. The current is limited by the maximum current capability of the regulator U4 (500 mA max).

Table 1-12 details the position of the JP5 jumper with regard to using an external power supply.

Using VIN or E5V as an External Power Supply

VIN or E5V can be used as an external power supply in case the current consumption of Nucleo-64 and the extension board exceeds the maximum allowed current provided by the USB connection. It is still possible to use the USB for programming and debugging only, but it is mandatory to power-on the Nucleo-64 board first using VIN or E5V and then connect the USB cable to the host PC. Following

Input Power Name	Connector Pins	Voltage Range	Limitations
3.3 V	CN6, pin 4; CN7, pins 12 and 16	3.0 to 3.6 V	Used when ST-LINK PCB is separated or SB2 and SB12 are off

Table 1-11 *3.3-V External Power Supply Configuration*

Jumper	Description
JP5	Jumped to positions 1 and 2 when the U5V (ST-LINK) is the power source
	Jumped to positions 2 and 3 when VIN or E5V is the power source

Table 1-12 *JP5 Jumper Positions*

this sequence ensures that the USB enumeration will occur when connected to an external power source.

The following power-on sequence procedure should be used:

1. Connect the jumper between pins 2 and 3 of JP5

2. Check that jumper JP1 is removed

3. Connect the external power source to VIN or E5V

4. Turn-on the external power supply VIN or E5V

5. Check that LD3 is turned on

6. Connect the host PC to USB connector, CN1

If this order is followed, the board may be supplied by VBUS first and then by VIN or E5V. The following risks may be encountered:

- The host PC USB power supply can be damaged if more than 300 mA is drawn by the Nuclueo-64 board and host cannot supply it.

- Up to 300 mA may be drawn from the host PC USB connection because JP1 is in the off position. If the host cannot supply this amount, the USB enumeration will fail and the MCU will not start.

This section concludes this introductory chapter for the STM32 Nucleo-64 board.

Summary

This is an introductory chapter on the STMicroelectronics (STM) Nucleo-64 microcontroller (MCU) development board. This board features the model STM32F302R8 MCU, which runs at 72-MHz clock rate and has 64-KB flash and 16-KB SRAM memories.

I first reviewed the differences between a microcomputer and a MCU. A review of the principal components that make up a MCU followed next. The first component discussed was the ARM Cortex M-4 32-bit microcomputer. The next was the memory, where I briefly explained how the Cortex M-4 Harvard Architecture speeds up the processing cycle. The last component portion was an extensive discussion of the MCU peripherals.

I carefully explained how the Arduino and ST Morpho connectors were set up on the board using a series of detailed tables with additional comments. Next followed some peripheral discussions concerning an interrupt controller, timers/counters, and bit-serial interfaces.

The chapter finished up with a thorough discussion of the many board options available with which you can customize the Nucleo-64 development board to suit your own situation.

2

STM MCU Software

This chapter explores the software side of MCU development. The discussion on MCU hardware in Chapter 1 becomes largely irrelevant without a thorough discussion on how to use it, which is the role of software. It would be helpful to provide a bit of prospective as it pertains to the software situation as related to STM MCUs.

Open-Source versus Commercial Proprietary Software

The first thing you need to realize is that STM is a multibillion dollar, for-profit company and must make a profit to stay in business as well as satisfy its stakeholders. Although STM does support the maker community and open-source development, it has fostered commercial software suites to be the primary means by which software is developed for their MCUs and other product lines. This means it is up to each individual maker and/or open-source developer to decide whether or not to craft your own open-source software toolchains or use a proprietary solution. I discuss both approaches and you will shortly see what my decision was and why I made that decision.

The available STM MCU commercial toolchains can be used with very little fuss but they are very expensive, with all of them costing more than $1,000 for a single license. Some software suppliers such as Keil and IAR do provide evaluation versions of their software freely available for those developers who want to try their suites out before purchasing. The only constraint with these versions is that they limit the uploaded MCU program size to 32 KB. While this limit might appear to be small, in reality, it is more than adequate for almost all projects the maker community would develop. I used the Keil evaluation IDE for all the book project code development. I have found that using this particular IDE makes developing code quite easy and with little issues that happen when attempting to develop using purely open-source tools. I want to be very clear that I am not recommending that you avoid the open-source approach, but simply stating that I have attempted both approaches and determined that using a proprietary IDE was much simpler and almost always worked the first time.

Bare Metal Development

The phrase "bare metal development" is a fairly recent addition to the developer's lexicography. It generally is taken to mean any low-level method of programming related to a specific hardware set. It is derived from the phrase "programming on the bare metal." This type of development has been around since the advent of MCUs in the 1970s, when it was the only type available. It wasn't called *bare metal development* at that time, just embedded development or simply MCU programming. It would be helpful to relate some history to understand the present state of MCU software development.

Brief History of MCU

The Texas Instruments (TI) TMS1000 is often called the *first microcontroller*. It was released in 1971 and had read-write memory (RAM), read-only memory (ROM), an embedded 8-bit processor, and an on-board clock system. Previous to this chip, other processors relied on external memory and clock peripherals for those functions. TMS1000 was used in early TI calculators as well as some gaming consoles. Meanwhile, Intel designed and produced Intel 8048, which too had on-board RAM and ROM, as well as an 8-bit processor, clock circuitry, and some GPIO pins for direct peripheral interface. This chip turned out to be immensely popular, especially with computer keyboard manufacturers. There were millions of 8048s used in keyboards and it soon became a big revenue generator for Intel.

TMS1000 and Intel 8048 were likely programmed using the C language along with a cross-compiler, which is a software application running on a host PC that produces raw binary code suitable for use with the embedded MCU processor. The early MCUs used a masked ROM to store the program. This meant the program was first developed, tested, and then converted to a mask that was used in

Figure 2-1   *Intel 8751 MCU.*

the MCU wafer production process. There was no ability to reconfigure a program in the field. As you might expect, the costs to create a masked MCU was very high and developers had to get it right the first time or risk very high retooling costs to change the program.

In the 1980s, MCUs started to appear with EPROMs, which stored programs that could be changed once the MCU was exposed to ultraviolet (UV) light. Figure 2-1 shows Intel 8751, which is UV reprogrammable version of the immensely popular 8051 MCU.

If look carefully at the figure, you may see the outline of a sticker that once was applied to cover the UV window. This covering was important to prevent ambient sunshine from inadvertently erasing the EPROM. The UV EPROM eventually gave way to the EEPROM, which is an electrically erasable ROM. The Intel 8052 is a MCU version that uses an EEPROM. This chip became hugely successful in the embedded development community with many millions, if not billions, being employed in a wide variety of systems, devices, and other embedded applications. In fact, Intel 8052 and certain variants are still being manufactured today. At one time, the 805x family was by far the world's most widely used MCU line. However, I now believe smartphone MCUs have overtaken the 805x's as the most widely used MCUs.

I include Table 2-1, which compares similar features between an 8052 and the STM32F302R8 MCUs used in this book. This table starkly shows huge improvements in technology in this field.

Just remember that in its day, the 8052 was considered a very powerful MCU. In the 1990s, I had the privilege of working with a new product development team that used an 8052 development board as its main controller. I programmed the MCU using the C language. Fortunately for the team and myself, we could reprogram it very easily as bugs in the product were discovered and resolved.

Feature	8052	STM32F302R8
Processor	8 bit	32 bit
Max clock rate	12 MHz	72 MHz
RAM	128 bytes	16 KB
ROM	4 KB	64 KB
Max memory address range	64 KB	4 GB
GPIO	4	51
Counter/timers	2, 16-bit counter/timers	9 counter/timers (mix of 16 and 32 bit)

Table 2-1   *Comparison between an 8052 and STM32F302R8 MCUs*

The MCU Toolchain

Embedded developers use a "toolchain" to create and debug software run on a MCU. There is no unique toolchain because what is used depends on the host computer being used, the target MCU, the development language, the linker/loader, and other various components. Generally speaking, the toolchain consists of an editor, cross-compiler, support libraries, linker/loader, and a debugger. Most of the toolchain components are often integrated into a single application called an *integrated development environment* (IDE) or they can be separate pieces for which the developer must connect and use in a proper sequence.

Using an IDE definitely helps a developer in efficiently creating code for a microcontroller because an IDE has the following:

- An integrated source code editor

- A means to create a software project and automatically create and store all necessary supporting files

- A navigation scheme to quickly sort through the many files that constitute a project

- One-step cross-compilation

- Easy integration with appropriate load and debug applications, which in this case is ST-LINK/V2

- Versatile debug modes

I also want to mention that the Keil IDE contains many more features than the Arduino IDE, which many readers already likely use. The Arduino IDE comes ready to use "right out-of-the box," which makes it very convenient for most users but at the same time limits its capabilities of supporting a vast variety of different microcontrollers, and you are limited to one programming language, that being, Processing. Setting up your own, customized toolchain simply provides you a very flexible way to support your unique programming requirements.

Toolchains may be quite simple, as is the case with the Arduino IDE, or very complex, as might be the situation for a unique MCU being developed in a Linux system with open-source software. I would consider the STM32 toolchain as established in this book as moderately complex. However, one piece of good news is that the manufacturer freely provides an application that nicely supports its MCU line as far as setting up clocks and peripherals using the C language. I discuss this application named STMCube32MX in a later section. Figure 2-2 shows the toolchain workflow for a generic embedded development project.

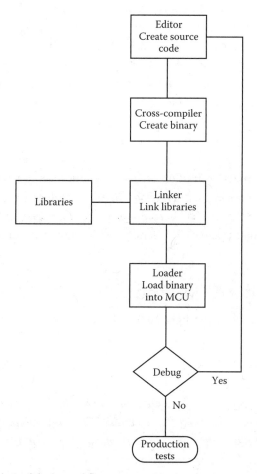

Figure 2-2   *Generic toolchain workflow.*

The first step in the process is to create the source code using a text editor. Most often, the editor is an integral part of the IDE, but developers often have their own favorite editor, which they have been accustomed to use. In all cases, the source is generated using the desired computer language, which will then be used as the input to a cross-compiler. The cross-compiler is so named because it functions with the host computer's operating system, yet produces binary code that will run on the target MCU. This newly generated binary code must then be linked with all necessary binary libraries to form a fully executable MCU program. That job is the responsibility of the linker. The fully formed and executable binary program must next be uploaded into the MCU. This is the responsibility of the loader. The loader function is often partitioned as part hardware and part software. The ST-LINK/V2 performs the load function for the Nucleo-64 development board used in this book's projects.

The MCU can now be run with a working program and tested to see if it performs as desired. This is called debugging and is shown as a diamond in the figure. If the program works without any flaws, which does not happen very often, the MCU is ready to go into the production or release phase. However, if bugs are detected, they must be resolved and this normally means iterating all the way back to the source code and making appropriate changes at that point. The whole toolchain must then be repeated as often as necessary until the MCU functions as desired or any left-over bugs are so trivial that they can be tolerated and stay in the final program executable code. One example of a trivial bug may be that the MCU responds to a sensor event in 21 ms instead of 20 ms, but that slight deviation has no measurable impact in the overall MCU performance.

Configuring a STM32 Toolchain

There are many ways to configure a toolchain, as previously mentioned. I have elected to use the Keil evaluation IDE, version 4.74.

NOTE   *I used a Windows 10 platform to host this toolchain. I did this because I determined that the ST-LINK/V2 utility worked properly with Windows as compared to using a Mac OS or a Linux environment. You can always choose to create a toolchain with a Mac or Linux host, but be prepared to encounter some compatibility problems when attempting to link and load programs from the host to the target platform.*

Installing the Keil IDE

The latest Keil IDE is version 5; however, I decided to install an older evaluation version 4.74 of the IDE because it seems to be more stable and I had more favorable experiences in using it. You may elect to install the latest version if you so decide, but all my projects will be done with the earlier version. The executable evaluation version 4.74 is available from the following website:

https://www.keil.com/demo/eval/armv4.htm

You will first have to fill out a registration form before you are allowed to access the download page. This form is shown in Figure 2-3 and doesn't appear to be too

Figure 2-3 *Keil registration form.*

onerous and just requests some basic information about yourself and how you intend to use the application.

You will see the download Web page, as shown in Figure 2-4, once you fill out the registration form and click on the Submit button.

The download file is named MDK474.exe and is a straight Windows executable of approximately 576 MB in size. I would also recommend that you just leave all the default installation parameters as they are because they worked just fine for my toolchain setup.

You should see the Keil IDE icon on the Desktop, as shown in Figure 2-5, once the installation is complete. Also notice that Keil brands its IDEs as uVision. You will also see the uVision tag on many files that are automatically created during the target build process. That's all that is needed to install the Keil IDE.

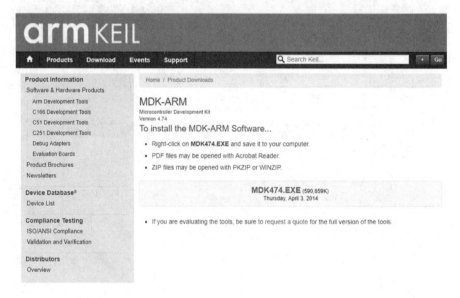

Figure 2-4   *IDE download Web page.*

Figure 2-5   *Keil IDE v4.74 icon.*

STM32CubeMX

The manufacturer STM provides a very nice application called STM32CubeMX that helps configure its line of MCUs and demonstration boards. The STM32CubeMX utility is a GUI-based tool that helps you create custom configuration files using the C language that are especially suited to the Nucleo board being used. For instance, this utility can create all the source code required to blink the user LED (LD2) on the NUCLEO-F302R8 board, which in turn uses a STM32F302R8 microcontroller. The C language source code covers the processor clock, peripheral port, and general-purpose input/output (GPIO) configuration as well as a few more. The STM32CubeMX application can be freely downloaded from the STM website, http://www.st.com/en/development-tools/stm32cubemx.html. You will have to register at the STM website in order to download this software, but it is free and STM will notify you of changes and updates to their tools as they are published. This application should be downloaded into the STM32Toolchain directory as was done with all the previous applications.

Figure 2-6 shows the opening splash screen for the STM32CubeMX utility after it has been downloaded and installed. I would recommend that you accept all the default suggestions as you run the installer application.

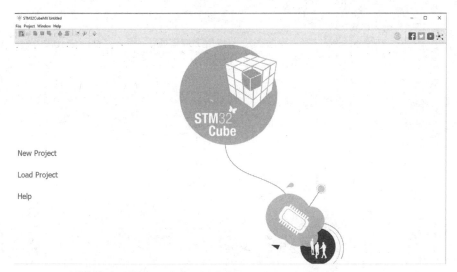

Figure 2-6 *STMCubeMX opening splash screen.*

I clicked on the "New Project" button in order to show you how this utility functions. A selection page will next appear with two tabs located near the upper left-hand corner of the window. These tabs are labeled as follows:

- MCU Selector

- Board Selector

I next selected the MCU Selector tab and scrolled down an extensive listing of microcontrollers until I encountered the STM32F302R8 MCU, which is the type installed on the Nucleo board used in this book. Figure 2-7 shows the screen with this particular MCU selected.

You should notice that a nice abstract of the MCU components and features appears in the window's upper-top section. This information is often useful to confirm that the MCU does support whatever functions you intend to implement.

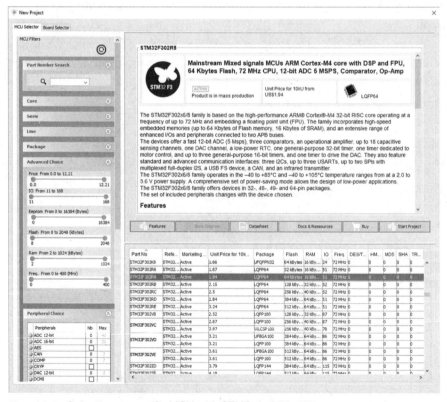

Figure 2-7 *Selecting a specific MCU with STMCubeMX utility.*

The other tab labeled Board Selector will be the one most useful to support a Nucleo board demonstration project. I clicked on that tab and scrolled down in order to select the precise Nucleo board used in the book, which is the Nucleo64 NUCLEO-F302R8 board. Figure 2-8 shows the result of making this selection.

A peripheral selection panel appropriate to the selected board appears near the window's left-hand side. You should notice that some peripherals will have any empty box appearing in the Nb column next to the peripheral name, which indicates that particular peripheral is not included on the selected board. There is also a Max column located next to the Nb column, which shows how many peripherals of a particular type are available for that specific Nucleo board. An additional handy feature is the ability to quickly download the selected Nucleo board user guide by clicking on the "Load User Manual" located near the window's lower right-hand corner.

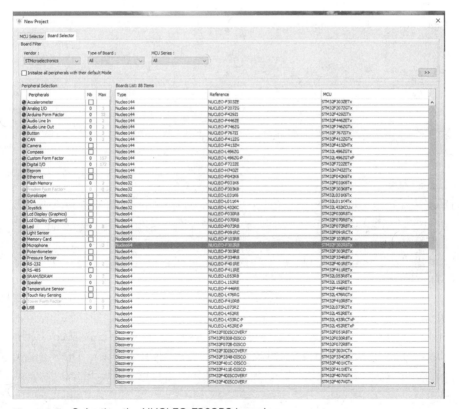

Figure 2-8 *Selecting the NUCLEO-F302R8 board.*

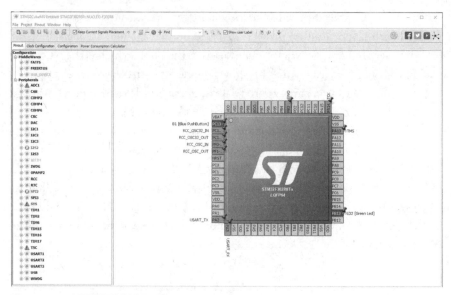

Figure 2-9 *STM32F302RE MCU pinout diagram.*

Figure 2-9 shows a very useful and convenient MCU pinout diagram for the unit installed on the Nucleo board. It shows all the pins already connected to various peripherals including the User LED, which I will demonstrate in the next chapter.

ST-LINK

The ST-LINK software is the next part of the toolchain, which must be installed in order to start developing the Nucleo board. It is an extremely important toolchain component because it supports the USB physical link between the board and the host computer and is the only means by which software can be uploaded to the board.

The USB ST-LINK driver software must first be downloaded and installed into the host computer in order to establish a communications link between the host and the Nucleo board. This driver software is available from the same st.com website where you downloaded the STM32CubeMX utility application. The software part number from the ST website is STSW-LINK009 and the technical description is "USB driver for ST-LINK/V2 and ST-LINK/V2-1." This particular download is a signed driver suitable for Windows 7, Windows 8, and Windows 10 for both 32- and 64-bit hosts. The downloaded driver is an archive format named "en-stsw-link009" and must be extracted from being run. Go into the newly created directory en-stsw-link009 and click on the Windows batch file named "stlink_winusb_install" to install the USB driver. You will briefly see a command window showing the installation in progress. There is no driver software icon created, so the only way to know if the driver installation successfully completed is to run the ST-LINK utility. The ST-LINK utility installation is discussed below.

The actual ST-LINK/V2 utility must be separately downloaded from the st.com website. Its part number is "STSW-LINK004" and it is approximately 22 MB in size. The downloadable archive is named "en.stsw-link004", which must also be extracted before being installed. Ensure that the archive has been downloaded into the STM32Toolchain directory. Next, go into the newly created directory en.stsw-link004 and click on the STM32-ST-LINK Utility V4.0.0.setup application, which will install and configure the ST-LINK/V2 utility. You should see the utility opening screen, as shown in Figure 2-10, after you click on the ST-LINK/V2 icon appearing on the host desktop.

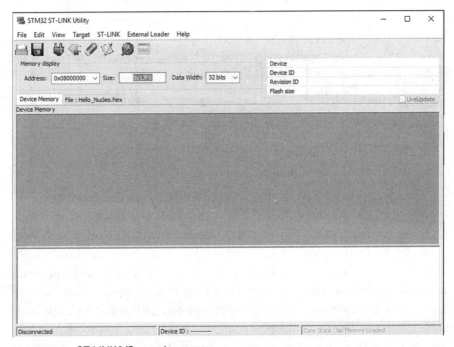

Figure 2-10 *ST-LINK/V2 opening screen.*

Figure 2-11 *USB cable.*

One of the first things you need to do is to check which ST-LINK firmware version is currently installed on the Nucleo board you are using. To do this, first ensure the ST-LINK utility is NOT running and then connect the Nucleo board to the host computer using a USB cable with mini-A connector that plugs into the Nucleo board and a standard type A connector that plugs into the host computer. Figure 2-11 shows the proper USB cable to be used.

Follow the steps below after connecting the Nucleo board to the host computer to determine the current firmware installed on the board.

1. Open the ST-LINK/V2 utility by clicking on the desktop icon.

2. Click on the "ST-LINK" selection located on the horizontal menu bar.

3. Click on "ST-LINK Upgrade" selection in the sub-menu.

4. Click on the "Device Connect" button.

Figure 2-12 shows the result after I clicked on the Device Connect button.

Obviously, do NOT upgrade the firmware version if the reported version is more current than the one shown as the upgrade. This odd situation likely happens because the upgrade version has been hard-coded into the ST-LINK/V2 archive, which in turn is not updated as often as the firmware in newly manufactured Nucleo boards. The way around this issue that companies often use is to have a dynamic download link in their upgrade software such that the latest firmware version is always available. Apparently, STM chose not to pursue this approach, which is why this situation happens. Hopefully, it will change its

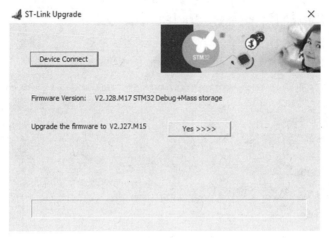

Figure 2-12 *ST-LINK firmware version display.*

upgrade technique in the near future. In any case, the firmware check, in itself, confirms that the USB communication link between the host and the Nucleo board is working properly.

The next step in the toolchain configuration is to set up a debugging tool named OpenOCD. It is not really required to install the debugger, but it may come in handy when you are trying to troubleshoot a program that doesn't work the way you believe it should.

OpenOCD15

The Open On-Chip Debugger (OpenOCD15) utility was designed to upload new firmware onto a STM32 Nucleo board as well as support an extensive software debugging mode. This utility was originally created by Dominic Rath, but is now maintained by the STM community, including the STM company. A Windows compatible version has been created by Liviu Ionescu and is available at https://github.com/gnu-mcu-eclipse/openocd/releases/tag/gae-0.10.0-20170124. Figure 2-13 shows the many OpenOCD15 versions that are available to be downloaded when you go to that website.

I selected the version named "gnuarmeclipse-openocd-win64-0.10.0-201701241841-setup.exe", which is the one applicable for use with a 64-bit Windows system. This executable file is 2.19 MB in size. Download this file to the STM32Toolchain directory so that it is readily available to any toolchain application that needs it. This file is a self-contained executable, which simply needs to

GNU ARM Eclipse OpenOCD v0.10.0-201701241841

 ilg-ul released this on Jan 24

Version v0.10.0-201701241841 is a major stable release. It includes binaries for Windows, macOS and GNU/Linux.

Continue reading »

 downloads 9k gae-0.10.0-20170124

Downloads

gnuarmeclipse-openocd-debian32-0.10.0-201701241841.md5	89 Bytes
gnuarmeclipse-openocd-debian32-0.10.0-201701241841.tgz	2.44 MB
gnuarmeclipse-openocd-debian64-0.10.0-201701241841.md5	89 Bytes
gnuarmeclipse-openocd-debian64-0.10.0-201701241841.tgz	2.4 MB
gnuarmeclipse-openocd-osx-0.10.0-201701241841.md5	83 Bytes
gnuarmeclipse-openocd-osx-0.10.0-201701241841.pkg	2.34 MB
gnuarmeclipse-openocd-win32-0.10.0-201701241841-setup.exe	2.15 MB
gnuarmeclipse-openocd-win32-0.10.0-201701241841-setup.md5	92 Bytes
gnuarmeclipse-openocd-win64-0.10.0-201701241841-setup.exe	2.19 MB
gnuarmeclipse-openocd-win64-0.10.0-201701241841-setup.md5	92 Bytes

Figure 2-13 *OpenOCD15 downloads.*

be run in order to install OpenOCD on your host computer. Figure 2-14 shows the opening splash screen for the software installation.

I recommend accepting all the default selections as the installation proceeds, which only took a minute or so for my particular situation. Figure 2-15 shows the completion screen after the OpenOCD setup software has finished its installation process.

Running the OpenOCD utility is a bit different than ordinary applications because it has been created to be a service or background program. In Unix and Linux terms, this program type is known as a daemon. There are two ways to access the program.

1. Using the telnet utility with port 4444: OpenOCD acts like a server when a telnet link is created. It will respond appropriately when commands are received. It is strictly a command-line environment.

2. Using the telnet utility as remote server for the GNU Project Debugger (GDB): OpenOCD implements the GDB remote protocol and may be used

Figure 2-14 *OpenOCD15 opening screen.*

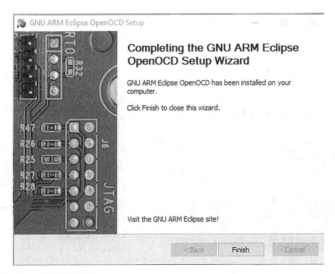

Figure 2-15 *OpenOCD finish screen.*

as a mediator component between GDB and the target platform. GDB uses the IDE's GUI environment.

I have elected to use the telnet approach because it is simple and direct and will serve most debugging purposes. GDB is more complex and is often coupled with the Eclipse IDE, which is not being used in this toolchain.

Establishing a Telnet Link

The very first thing that you must do prior to using telnet is to enable it on Windows 10. Unfortunately, it is not enabled by default and must be done by the user. It is a relatively easy procedure as detailed below:

1. Enter "Control Panel" in the Search text box

2. Click on "Programs"

3. Click on "Program and Features"

4. Click on "Turn Windows features on or off" found in the left-hand column

5. Scroll down and click on the check box next to "Telnet Client"

6. Click on the "OK" button

There is an equivalent command-line procedure to enable telnet, but I found the GUI approach just as effective.

You will next need to determine the appropriate telnet script that supports the particular board that is being used for your projects. All the available scripts currently supporting Nucleo boards are detailed in Table 2-2.

I used the configuration file named "st_nucleo_f3.cfg" because it was associated to my NUCLEO-F302R8 board.

Nucleo Board Number	OpenOCD 0.10.0 Board Script File
NUCLEO-F446RE	st_nucleo_f4.cfg
NUCLEO-F411RE	st_nucleo_f4.cfg
NUCLEO-F410RB	st_nucleo_f4.cfg
NUCLEO-F401RE	st_nucleo_f4.cfg
NUCLEO-F334R8	stm32f334discovery.cfg
NUCLEO-F303RE	st_nucleo_f3.cfg
NUCLEO-F302R8	st_nucleo_f3.cfg
NUCLEO-F103RB	st_nucleo_f103rb.cfg
NUCLEO-F091RC	st_nucleo_f0.cfg
NUCLEO-F072RB	st_nucleo_f0.cfg
NUCLEO-F070RB	st_nucleo_f0.cfg
NUCLEO-F030R8	st_nucleo_f0.cfg
NUCLEO-L476RG	st_nucleo_l476rg.cfg
NUCLEO-L152RE	st_nucleo_l1.cfg
NUCLEO-L073RZ	stm32l0discovery.cfg
NUCLEO-L053R8	stm32l0discovery.cfg

Table 2-2   *Telnet Scripts*

The next part of this process is to open a terminal window. The easiest and fastest way is to enter "cmd" in the Search textbox. Type in the following two command lines into the window:

```
cd c:\STM32Toolchain\openocd\scripts
..\bin\openocd.exe -f board\st_nucleo_f3.cfg
```

Remember to use the appropriate configuration file associated with your Nucleo board.

The OpenOCD service should then respond in the terminal window with what is shown in Figure 2-16.

Notice that an error message is shown in the OpenOCD response sequence, but I found that it was inconsequential and did not affect the program operations. However, if you do not see the responses following the error, it means that a real problem exists because no communications link was created between the OpenOCD program and the target board. This problem is usually caused by having an incorrect libusb version trying to be used by Windows. Follow the Zadig utility procedure to correct this issue.

1. Download the Zadig utility from http://zadig.akeo.ie/ for your Windows version.

2. Connect the Nucleo board to your computer.

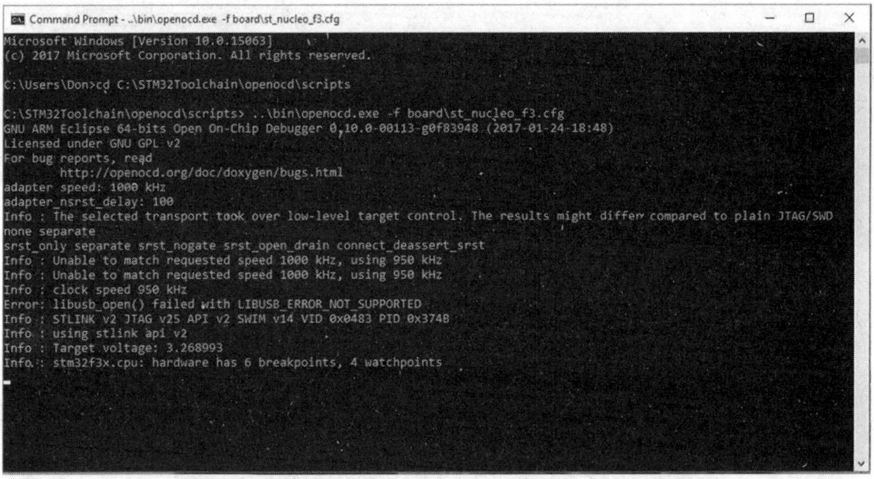

Figure 2-16 *Starting the OpenOCD service.*

3. Run the Zadig utility and select Options → List All Devices.

4. The Nucleo board should appear in the device list combo box after a short interval.

5. Select the WinUSB driver if it is not already listed.

6. Click on the Reinstall Driver button as shown in Figure 2-17.

This action should fix the libusb driver issue. Go back and try starting the OpenOCD service as previously described.

You can now minimize the terminal window, but do not close it because the OpenOCD service will stop if you do. Open another terminal window and enter the following command:

```
telnet localhost 4444
```

The OpenOCD service should respond with message as shown in Figure 2-18.

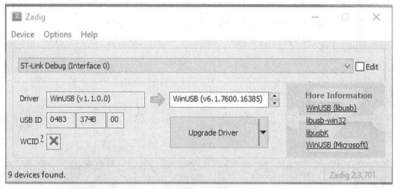

Figure 2-17   *Reinstall WinUSB driver dialog box.*

Figure 2-18   *Opening OpenOCD response.*

There will be a prompt awaiting your input to the service. There are many ways of interacting with OpenOCD and typing in "help" will give a big response for all of the many varied commands. Figure 2-19 shows just the beginning portion of the help response.

For much greater details, I would highly recommend that you download the OpenOCD user guide in PDF format from http://openocd.org/doc/pdf/openocd.pdf.

Figure 2-19 *Beginning portion of the OpenOCD help request.*

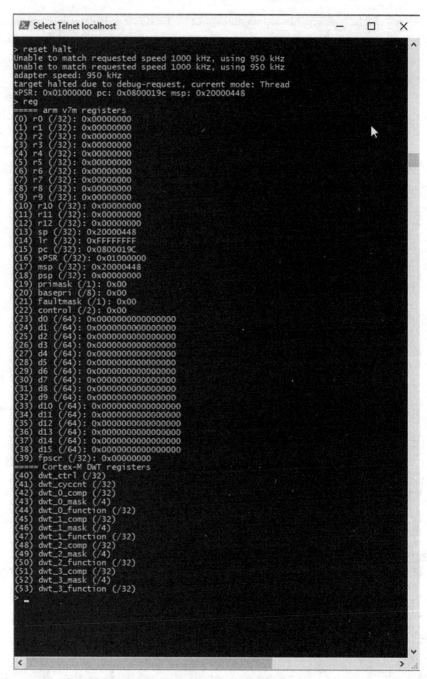

Figure 2-20 *OpenOCD reg command results after a reset.*

This is a 172-page document and currently references version 0.10.0 as of January 15, 2017.

I did try out a few commands just to see how OpenOCD worked. The first command was to halt the Nucleo board from running a program. This command was:

```
reset halt
```

The board was running a "blinky" type program when the command was given. It instantly stopped after I pressed the Enter key on the keyboard. The next command displayed the current contents of all the MCU registers. This command was simply:

```
reg
```

Figure 2-20 shows the impressive results of using this very simple command.

It is obvious that all the data registers were reset to 0 content based on the `reset` command that was given previous to the `reg` command. In Figure 2-21, I halted the blinky program without issuing a reset. In this case, the contents of the general-purpose registers were not reset as you can see in the figure.

OpenOCD also has a very handy command to examine any memory location. The location contents can be returned as a word, half-word, or byte using the

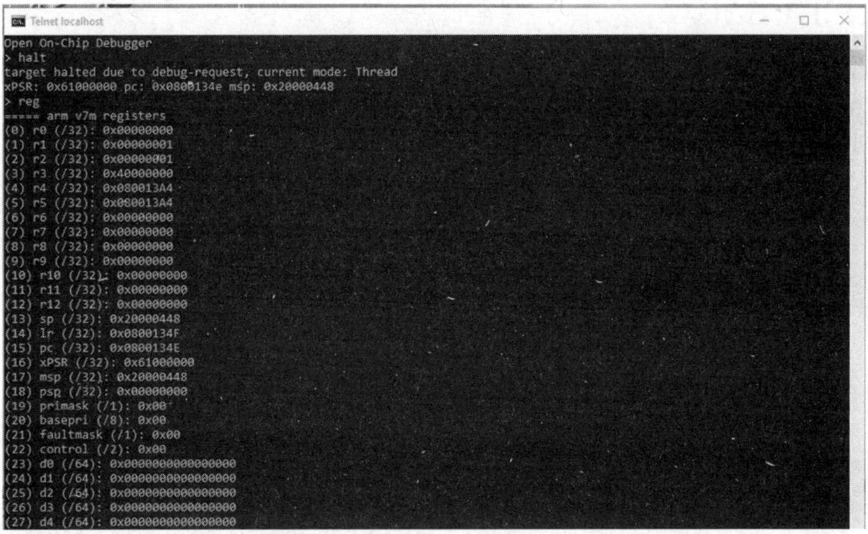

Figure 2-21 *OpenOCD reg command results after a halt.*

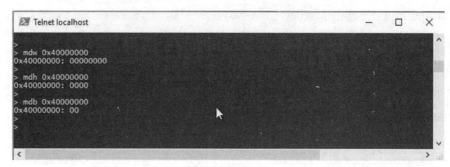

Figure 2-22 *Memory content display commands.*

commands mdw, mdh, and mdb, respectively. Figure 2-22 shows the results of these commands being executed on memory location 0x4000 0000.

I could expand upon the OpenOCD commands, but I feel you probably understand by now that the only way to get comfortable with OpenOCD is to use it and explore its many features.

The OpenOCD installation completes the overall toolchain installation process. I will demonstrate in the next chapter how to create a new C language project using the toolchain established in this chapter.

Summary

This chapter's discussions provide instructions on how to set up a STM32 toolchain that will enable you to create software projects that will execute on STM MCUs and demonstration boards. Instructions were only provided for Windows-based host platforms. However, I did include a reference that shows how to set up toolchains on Mac OSX and Linux platforms.

The Keil IDE using the C/C++ languages is used for this particular toolchain.

3

STM32CubeMX
Application

In this chapter I present an in-depth discussion of the STM32CubeMX application, which is a free utility provided by STMicroelectronics (STM). I will also start referring to it as CubeMX to just simplify the terminology. I would also like to acknowledge a fine Leanpub eBook published by Carmine Noviello titled *Mastering STM32*. Carmine's book was invaluable reference alongside the STM datasheets without which I could not have created this book.

CubeMX is a fairly complex utility because it is intended to encompass nearly all the microcontroller chips (MCUs) and development boards that STM offers. This is quite an amazing feat considering that there are hundreds of MCU variants and dozens of development boards that STM manufactures. The CubeMX supports these devices by automatically generating C code that developers can easily use in their programs. The generated code is designed to support specific integrated peripheral (IP) devices included on the selected MCU. CubeMX may be regarded as a MCU-centric utility because of the following features:

- MCU selection based on the STM family, that is, F0, F1, ...
- Specific IP required for the project involving mapping MCU peripherals to a specific pinout
- Handles specific MCU clock, power, interrupt control, ...
- Automatically adapts pinout to package configuration
- Manages the ST hardware abstraction layer (ST HAL)

- Manages external libraries

- Adaptable to a variety of IDEs

Figure 3-1 shows the opening screen when the application is initially launched. There are three choices available on the screen:

- New Project

- Load Project

- Help

The New Project selection will be shortly discussed in-depth, but I do first want to briefly mention what happens with the other two selections. Clicking on Load Project will display the open file dialog box, as shown in Figure 3-2.

In this figure there is only one project file named Hello.ioc, which is located in a project directory that is automatically created whenever a new project is created. The project directory is named using the same name entered in the new project dialog box. The project file contains everything that CubeMX needs to reconstitute the project in order to continue with any development efforts.

Clicking on the Help selection will display the opening page of a PDF document, as shown in Figure 3-3.

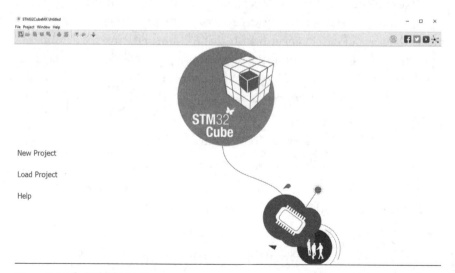

Figure 3-1 *CubeMX opening screen.*

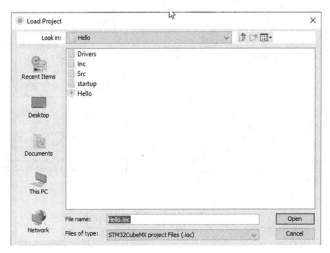

Figure 3-2   *Load Project screen shot.*

This PDF revision is a 261-page document that carefully details all the many features and functions contained within the CubeMX application. I would strongly recommend that you select and open this document and then save it to any convenient directory. In that way, you will always have access to the help document without needing to open the CubeMX application itself. You should also consider printing applicable portions of the document to have ready reference to appropriate material while using the application.

Clicking on the New Project selection will display a page with two tab selections. One tab is entitled MCU Selector and the other is Board Selector. Clicking on the Board Selector will cause the dialog box shown in Figure 3-4 to appear.

I selected a Nucleo64 as the board type and NUCLEO-F302R8 as the reference as may be seen in the figure, which matched the demonstration board I selected to use in this book. Figure 3-5 shows the result of selecting this particular board.

There are four tabs shown in this figure, which I will discuss here one by one. These tabs, displayed from left to right, are as follows:

- Pinout

- Clock Configuration

- Configuration

- Power Consumption Calculator

 lile.augmented

UM1718
User manual

STM32CubeMX for STM32 configuration and initialization C code generation

Introduction

STM32CubeMX is a graphical tool for 32-bit ARM® Cortex® STM32 microcontrollers. It is part of STMCube™ initiative (see *Section 1*) and is available either as a standalone application or as an Eclipse plug-in for integration in Integrated Development Environments (IDEs).

STM32CubeMX has the following key features:

- Easy microcontroller selection covering whole STM32 portfolio
- Board selection from a list of STMicroelectronics boards
- Easy microcontroller configuration (pins, clock tree, peripherals, middleware) and generation of the corresponding initialization C code
- Easy switching to another microcontroller by importing a previously-saved configuration to a new MCU project
- Easy exporting of current configuration to a compatible MCU
- Generation of configuration reports
- Generation of embedded C projects for a selection of integrated development environment tool chains. STM32CubeMX projects include the generated initialization C code, MISRA 2004 compliant STM32 HAL drivers, the middleware stacks required for the user configuration, and all the relevant files for opening and building the project in the selected IDE.
- Power consumption calculation for a user-defined application sequence
- Self-updates allowing the user to keep the STM32CubeMX up-to-date
- Download and update of STM32Cube embedded software required for user application development (see *Appendix E: STM32Cube embedded software packages* for details on STM32Cube embedded software offer)

Although STM32CubeMX offers a user interface and generates a C code compliant with STM32 MCU design and firmware solutions, it is recommended to refer to the product technical documentation for details on actual implementation of microcontroller peripherals and firmware.

The following documents are available from *http://www.st.com*:

- STM32 microcontroller reference manuals and datasheets

STM32Cube HAL driver user manuals for STM32F0 (UM1785), STM32F1 (UM1850), STM32F2 (UM1940), STM32F3 (UM1786), STM32F4 (UM1725), STM32F7 (UM1905), STM32L0 (UM1749), STM32L1 (UM1816) and STM32L4 (UM1884).

Figure 3-3   *Opening page for CubeMX Help document.*

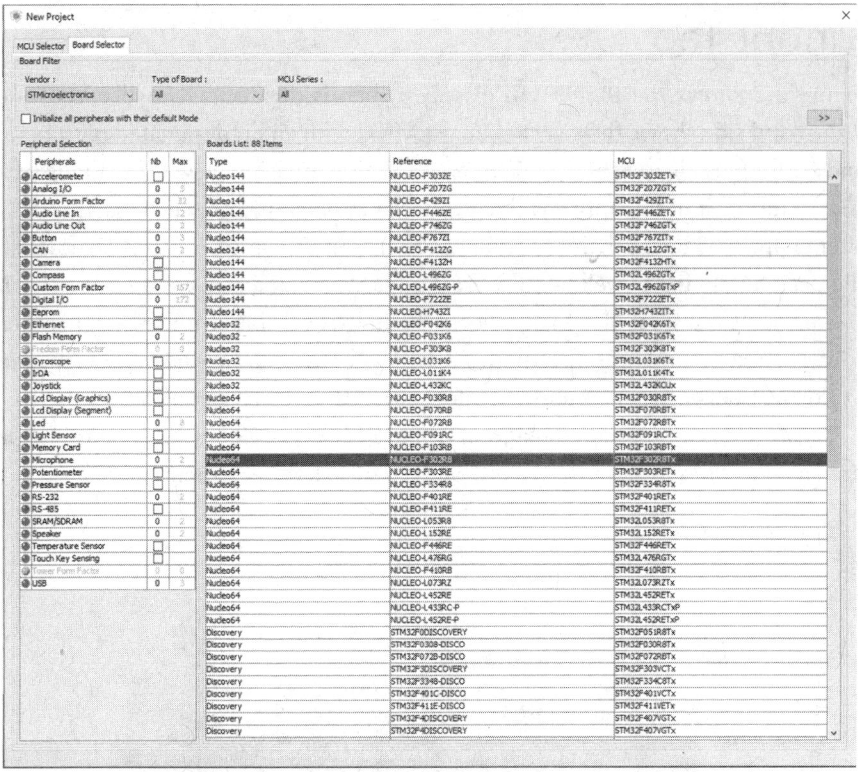

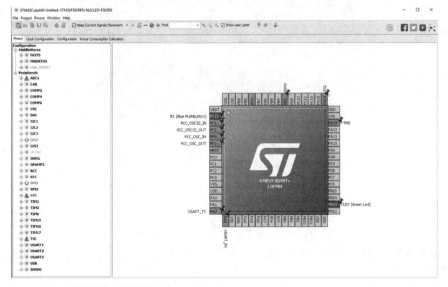

Figure 3-4 *Board Selector dialog screen shot.*

Figure 3-5 *NUCLEO-F302R8 board selection.*

Pinout Tab

Figure 3-5 shows the Pinout tab display, which is divided into two parts. The right-hand side shows the selected board MCU with all predesignated peripheral and GPIO pins clearly shown. ST refers to this as the Chip view for obvious reasons. These pin designations will be explored in detail in later sections as well as later chapters. The left-hand portion of the Pinout display in Figure 3-5 shows all the peripheral pins as well as software MiddleWares in a tree view. Next to each tree element is a symbol, which reflects the individual status of the peripheral or software line item. I will discuss these symbols after discussing Chip view and exploring the significant amount of information that can be gained from this view.

Figure 3-6 is an expanded diagram of the STM32F302R8Tx MCU chip that is used as the core processor in the NUCLEO-F411RE development board.

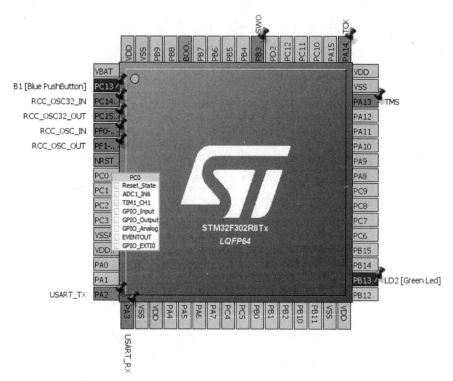

Figure 3-6 *STM32F302R8Tx MCU.*

NOTE   *It will be somewhat difficult for readers of the hard copy book to follow the next discussion as the figure uses color coded pins, which are very hard to distinguish in the gray-scaled rendered images. Therefore, I will attempt to identify the pins under discussion as completely as I can, given this constraint.*

There are two fully enabled GPIO pins colored in green, which are PC13 and PB13. PC13 is located near the top left-hand corner, while PB13 is at the lower right-hand corner. Pin PC13 is permanently connected to the development board's user blue push button and PB13 is permanently connected to the user green LED. These connections are also noted in the diagram, as shown in Figure 3-6.

There are nine orange-coded pins located on all four sides of the board. An orange-coded pin indicates that a peripheral normally assigned to a specific pin has not been enabled. For example, PA2 and PA3 are located on either side of the lower left-hand corner of Figure 3-6 and have the peripheral USART TX and RX functions assigned to each pin, respectively. Being orange coded will mean that the CubeMX application will not automatically generate any C code that will enable the USART functions for these pins.

All of the yellow-coded pins are typically located at or near the corners of the chip and are solely dedicated to power source functions with designations such as V_{dd}, V_{ss}, V_{bat}, and V_{ca}. These pins are strict hardware designations and cannot be programmed or reconfigured.

Finally, there are two remaining olive colored pins, BOOT and NRST, with BOOT located on the top edge and NRST located on the left edge. Like the power source pins, these two pins have dedicated MCU functions and cannot be programmed or reconfigured.

All remaining pins in the Chip view are gray-coded GPIO pins and can be initialized to suit program requirements. How this is accomplished is thoroughly discussed in the next chapter. All these pins have various functions or status outputs, which may be revealed by hovering the mouse over a particular pin. The functions and status outputs for pin PC0 as shown in Figure 3-6 are as follows:

- Reset_State
- ADC1_IN6
- TIM1_CH1
- GPIO_Input
- GPIO_Output

* GPIO_Analog

* EVENTOUT

* GPIO_EXTI0

It would be difficult at this point to explain each of these functions and status outputs. They are all discussed in the comprehensive STM32F302R8 MCU datasheet. The GPIO functions, however, will be discussed in the next chapter.

MCU Alternative Functions

The STM MCU architecture allows for alternative pin mapping for various peripheral functions. Control clicking on any enabled GPIO pin such as PC13 will reveal the alternate pin mappings associated with that pin, as shown in Figure 3-7.

All available alternative pins will immediately be color-coded blue in response to control clicking on a selected pin. In Figure 3-7, the alternative pins are PA13 and PB13, but only if the alternative pin(s) are not in a reset state. Using alternative

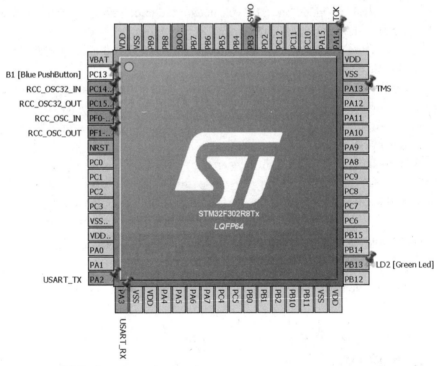

Figure 3-7 *PC13 alternative pin mappings.*

mapping allows for optimized board layout if it is inconvenient to layout a trace to the primary pin. The CubeMX will attempt to automatically resolve pin layout conflicts whenever detected by assigning a peripheral function to an alternative pin. However, by placing a logical "push pin" on a selected MCU pin will lock a desired peripheral function to a pin. These "push pins" are easily seen in the figure. These pins prevent CubeMX from enabling an alternative mapping of the locked pins. Pins locked out from alternative mapping are also color coded orange.

Integrated Peripheral (IP) Tree Pane

The IP tree pane shown in the left-hand side of Figure 3-6 allows a CubeMX user an easy way to enable or disable peripherals that may be assigned to a MCU pin. The tree pane also allows the user to enable certain software middleware libraries. The tree listing uses a variety of colored icons and text to indicate the peripheral status and configuration. Figure 3-8 shows these colored icons and textual

Display	Peripheral status
CAN1	The peripheral is not configured (no mode is set) and all modes are available.
ADC1	The peripheral is configured (at least one mode is set) and all other modes are available.
⚠ ADC3	The peripheral is configured (one mode is set) and at least one of its other modes is unavailable.
⚠ ADC2	The peripheral is not configured (no mode is set) and at least one of its modes is unavailable.
ETH	The peripheral is not configured (no mode is set) and no mode is available. Move the mouse over the peripheral name to display the tooltip describing the conflict.
CAN1 Mode Disable	Available peripheral mode configurations are shown in plain black.
⚠ DCMI DCMI Disable DMA ETH Slave-8-bits-Embedded-Synchro FSMC Slave-8-bits-External-Synchro I2C1 Slave-10-bits-External-Synchro I2C2 Slave-12-bits-External-Synchro I2C3 Slave-14-bits-External-Synchro	The warning yellow icon indicates that at least one mode configuration is no longer available.
ETH Mode Disable	When no more configurations are left for a given peripheral mode, this peripheral is highlighted in red.
LWIP: LightWeight TCP/IP Not available: Active only with ETH IP configured	Some modes depends on the configuration of other peripherals or middleware modes. A tooltip explains the dependencies when the conditions are not fulfilled.

Figure 3-8 *Tree pane IP icons and textual descriptions.*

meanings associated with each of them. This figure is excerpted from the CubeMX user manual, which I urge interested readers to download and save. There is a wealth of information available in the 261-page PDF manual, which I simply do not have the space to cover.

Clock Configuration Tab

The STM32 series of MCUs use a very flexible approach to configure a variety of clock sources needed for MCU operations. These sources apply to both core MCU functions and most of the peripheral functions. Clock sources and the associated phase-locked loops (PLL) are shown graphically in Figure 3-9, which will be displayed when you click on the Clock Configuration tab.

Figure 3-9 appears a bit daunting at first glance, but do not despair as it is really not that complicated, especially after you interact with it a few times. The STM32 clock sourcing is very much capable and flexible as compared to ordinary 8-bit MCUs.

The very first thing that must be considered is the basic clock sources. There is a high-speed clock (HSE) and a low-speed clock (LSE) used for the STM32 MCU series. These clocks may be sourced either internally or externally. These clocks

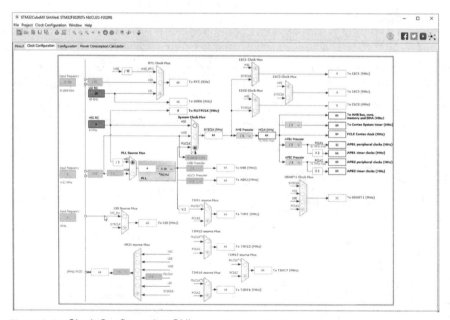

Figure 3-9 *Clock Configuration GUI.*

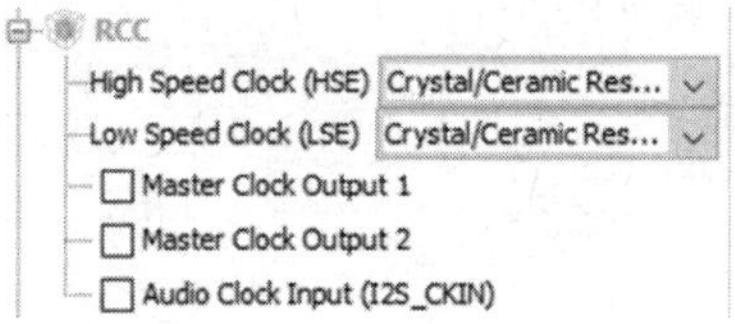

Figure 3-10   *RCC pin configurations.*

will be internally sourced for all of this book's projects; however, either one or both can be externally sourced if so desired. You must first enable the appropriate clock input pins if it desired to use external clock(s). This is done by returning to the IP tree in the Pinout view and expand the RCC section. Figure 3-10 shows a typical configuration for these pins.

You must return to the Clock Configuration tab once the clock input pins have been enabled and complete the clock source configurations.

There will be no need to change any of the clock configurations because they are perfectly suitable to support all book projects as they are set in their default states. I do want to caution readers who might be considering overclocking the MCU. Overclocking is a common hack that users will sometimes attempt in order to "gain" some perceived performance increase from a computer system. Overclocking is especially prevalent in gaming systems where high clock rates are thought to improve display frame rates and game performance. While this may be true for some gaming systems, the undue stress due to much higher chip heat dissipation will shorten processor life considerably. Gaming systems also employ sophisticated cooling subsystems, which are not present in the STM Nucleo board series. The bottom line is to NOT overclock as that will definitely overheat the MCU and cause it to prematurely fail.

Configuration View Tab

The Configuration View provides an overview for all of the software configurable components including GPIOs, peripherals, and middleware. There are clickable buttons, as can be seen in Figure 3-11, which allow for configuration options for various components. The CubeMX application will automatically create the supporting C/C++ code that supports the desired configuration.

There are color-coded icons and text in Figure 3-11, in much similar fashion to what was displayed in the IP tree pane view. Figure 3-12 details what the icons and text mean as they are applied to the Configuration View.

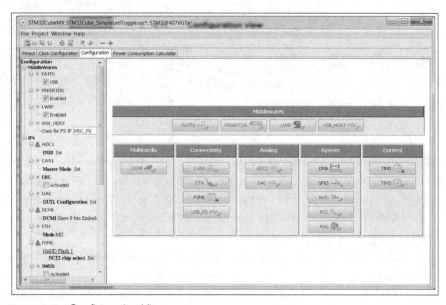

Figure 3-11   *Configuration View.*

Format	Peripheral Instance configuration status
DMA	Available but not fully configured yet. Click to open the configuration window.
ETH	Well configured with default or user-defined settings that allows proceeding with the generation of corresponding initialization C code. Click to open the configuration window.
TIM4	Badly configured with some wrong parameter values. Click to display the errors highlighted in red. Other example (UART): Baud Rate ✖ 1000000 Bits/s
TIM4 Configuration ⚠ Please select a Trigger Source in the Pinout view OK	Dialog box that explains source of error. It shall be fixed in another view.

Figure 3-12   *Configuration View icons and textual descriptions.*

Power Consumption Calculator View Tab

Clicking on the remaining tab will display an uninitialized power consumption calculator screen, as shown in Figure 3-13.

This PCC will allow you to calculate the following:

* Average current consumption

* Expected battery life

* Average DMIPS (Dhrystone million instructions per second): It is essentially a measure of the MCU efficiency as computed using a standardized software test

* Maximum ambient temperature (TAMAX): The maximum ambient temperature is computed based on the MCU package type, internal power consumption, and a preset 105°C maximum junction temperature

There are a series of procedural steps that are carefully detailed in the CubeMX user manual, which will lead you through a step-by-step analysis to ensure that the selected battery will handle your desired application. I will not use the PCC because all the book projects use a USB power source.

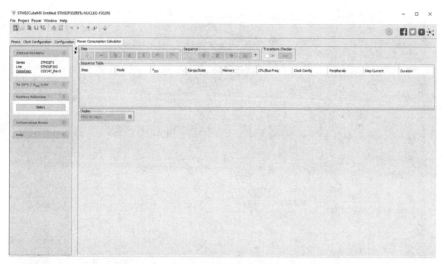

Figure 3-13   *Power consumption calculator (PCC).*

Creating an Example Project using CubeMX

It is now time to discuss how to create a new project using the CubeMX application. First start the application and then click on The New Project selection. The opening New Project screen, as shown in Figure 3-14, will be displayed.

Click on the Board selection tab, which should cause the Board Selection screen to be shown. Select your project board as I did in Figure 3-4. Double-clicking on the board selection will cause a Pinout window view to appear for the STM32F-302R8Tx MCU, as shown in Figure 3-5.

You will now need to click on the menu bar selection title Project in order to name your new project. Select Setting from the drop-down menu, which is shown in Figure 3-15.

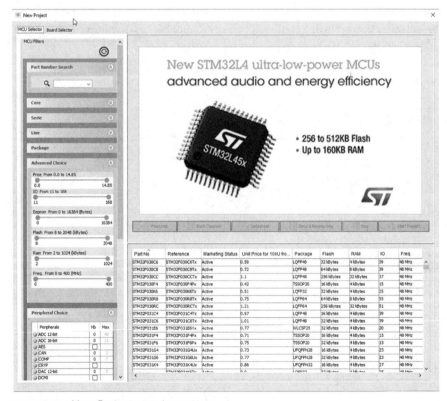

Figure 3-14 *New Project opening screen.*

Figure 3-16 shows the Project Settings window that will appear after you click on the Settings selection in the Project dropdown menu.

You must first enter a name in the Project Name textbox. I used "STM32Nucleo_F302" as a project name, but you can choose anything that you like. You should ensure that the Project Location textbox displays where you desire the project

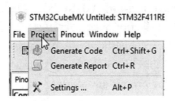

Figure 3-15 *Project dropdown menu.*

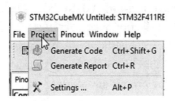

Figure 3-16 *Project Settings screen.*

folder to be stored. You can use the Browse button to find the appropriate directory. For this demonstration, I used the same location I previously used for earlier book projects. The next textbox entitled *Toolchain Folder Location* should be a directory named using the combination of the Project Location and the Project Name. Make note of this location because you will shortly need to access files that CubeMX will automatically create and store in it.

Next, ensure that you have selected "MDK-V4" as the Toolchain/IDE to use, as shown in Figure 3-16. This selection is very important because CubeMX cannot generate the correct C/C++ code without this being selected. In addition, ensure that the Generate Under Root checkbox is checked, which will allow the generated code to be located in the proper directories.

The default linker Settings are okay. Check that the MCU Reference textbox matches the MCU you are using. If it does not match, then the generated code cannot function with the physical MCU.

Finally, the Firmware Pack Name and Version textbox as well as the Use Default Firm Location textbox should be okay as is.

In preparation for code generation, go back and click on the Project menu bar selection and click on the Project Settings submenu selection. Next, click on the Code Generator tab and the screen shown in Figure 3-17 should appear.

Ensure that all the radio buttons and checkboxes are selected, as shown in Figure 3-17. You should note that I selected the radio button to "Copy only the necessary library files" in order to minimize the size of the project as stored on the host computer. Selecting the radio button "Copy all used libraries into the project folder" would have increased the overall project folder size. In either case, the final binary file size would still be the same.

Next, return to the Project submenu as has been shown in Figure 3-17, but this time click on the Generate Code selection. The CubeMX will take a few seconds to generate the code with a progress bar appearing on the screen. When the code generation is completed, you should see the informational dialog box, as shown in Figure 3-18.

After you click the Open Project button, you have the Keil IDE main window with the STM32Nucleo_F302 project directories and files listing appearing in the IDE's Project pane. This view is shown in Figure 3-19.

There should be four subdirectories in the Project pane named STM32Nucleo_ F302, Application/MDK-ARM, Drivers/CMSIS, and Application/User, respectively. In Figure 3-20, I expand all the subdirectories to show you all the files contained in them.

Project Settings   ✕

Project Code Generator Advanced Settings

STM32Cube Firmware Library Package

○ Copy all used libraries into the project folder

◉ Copy only the necessary library files

○ Add necessary library files as reference in the toolchain project configuration file

Generated files

☐ Generate peripheral initialization as a pair of '.c/.h' files per peripheral

☐ Backup previously generated files when re-generating

☑ Keep User Code when re-gene___ Before re-generating, all sources & header files are saved with '.bak' extension

☑ Delete previously generated files when not re-generated

HAL Settings

☐ Set all free pins as analog (to optimize the power consumption)

☐ Enable Full Assert

Template Settings

Select a template to generate customized code   Settings...

                                              Ok   Cancel

Figure 3-17   *Code Generator screen.*

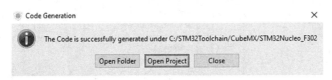

Code Generation   ✕

ⓘ The Code is successfully generated under C:/STM32Toolchain/CubeMX/STM32Nucleo_F302

              Open Folder   Open Project   Close

Figure 3-18   *Successful code generation dialog box.*

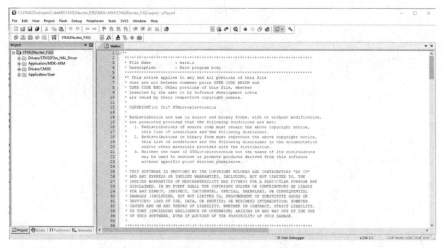

Figure 3-19 *Initial STM32Nucleo_F302 project in the Keil IDE.*

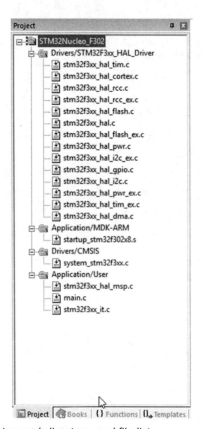

Figure 3-20 *Project directory, subdirectory, and file list.*

Your listing should be similar, depending upon the particular MCU that you used. Don't be dismayed by the number of directories and files. You will only be dealing with three at most in the upcoming chapter projects.

There will also be a pop-up dialog box warning you that a device was not found. This dialog box is shown in Figure 3-21 and is simply a consequence that the initial project load did not have a specific MCU selected.

This is easily fixed by clicking on File → Device Database, as shown in Figure 3-22.

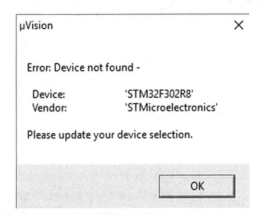

Figure 3-21 *Device not found dialog box.*

Figure 3-22 *Device Database menu selection.*

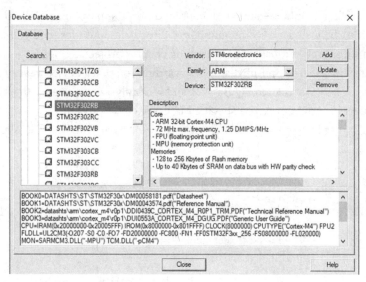

Figure 3-23   *STM32F302RB selection.*

Next select STMicroelectronics as the manufacturer and scroll down its rather impressive list of MCUs until you find your specific model. In my case, I did not find STM32F302R8, but did find STM32F302RB, which seems to work quite well. Maybe there will be some problems in the future, but I have not encountered any to date with this selection. Figure 3-23 shows the result of the STM32F302RB selection.

Just click on the Close button because there is not a Submit or Select button.

The main.c Code Listing

The following code listing is a file named main.c which was automatically generated by the CubeMX application based solely on the parameters and configuration settings that I entered. My purpose for this listing is to show you the basic template, which will constitute the bulk of the code for all the remaining projects in this book. In this listing is a method named main, which is where the majority of any functional code will be placed. In this template, it is intentionally empty because there is no functionality desired for this program. I am also required to show the voluminous STM disclaimer comments, which are part of STM's copyright requirements to use this code. This disclaimer will be present in all the actual code examples presented in the book. However, it will not be expanded as shown in this listing. You can always click on a small + sign to display the disclaimer.

After the listing, I have provided some additional comments regarding the methods contained in the listing.

```
/**
  * File Name        : main.c
  * Description      : Main program body
  * This notice applies to any and all portions of this file
  * that are not between comment pairs USER CODE BEGIN and
  * USER CODE END. Other portions of this file, whether
  * inserted by the user or by software development tools
  * are owned by their respective copyright owners.
  *
  * COPYRIGHT(c) 2017 STMicroelectronics
  *
  * Redistribution and use in source and binary forms, with or without
  * modification, are permitted provided that the following conditions
  * are met:
  *   1. Redistributions of source code must retain the above
  *      copyright notice, this list of conditions and the following
  *      disclaimer.
  *   2. Redistributions in binary form must reproduce the above
  *      copyright notice, this list of conditions and the following
  *      disclaimer in the documentation and/or other materials
  *      provided with the distribution.
  *   3. Neither the name of STMicroelectronics nor the names of its
  *      contributors may be used to endorse or promote products
  *      derived from this software without specific prior written
  *      permission.
  *
  * THIS SOFTWARE IS PROVIDED BY THE COPYRIGHT HOLDERS AND
  * CONTRIBUTORS "AS IS" AND ANY EXPRESS OR IMPLIED WARRANTIES,
  * INCLUDING, BUT NOT LIMITED TO, THE IMPLIED WARRANTIES OF
  * MERCHANTABILITY AND FITNESS FOR A PARTICULAR PURPOSE ARE
  * DISCLAIMED. IN NO EVENT SHALL THE COPYRIGHT HOLDER OR CONTRIBUTORS
  * BE LIABLE FOR ANY DIRECT, INDIRECT, INCIDENTAL, SPECIAL,
  * EXEMPLARY, OR CONSEQUENTIAL DAMAGES (INCLUDING, BUT NOT LIMITED
  * TO, PROCUREMENT OF SUBSTITUTE GOODS ORSERVICES; LOSS OF USE, DATA,
  * OR PROFITS; OR BUSINESS INTERRUPTION) HOWEVER CAUSED AND ON ANY
  * THEORY OF LIABILITY, WHETHER IN CONTRACT, STRICT LIABILITY,
  * OR TORT (INCLUDING NEGLIGENCE OR OTHERWISE) ARISING IN ANY WAY OUT
  * OF THE USE OF THIS SOFTWARE, EVEN IF ADVISED OF THE POSSIBILITY OF
  * SUCH DAMAGE.
  *
  */

// Includes
#include "main.h"
#include "stm32f3xx_hal.h"

/* USER CODE BEGIN Includes */

/* USER CODE END Includes */
```

```
// Private variables

/* USER CODE BEGIN PV */
// Private variables

/* USER CODE END PV */

// Private function prototypes
void SystemClock_Config(void);
static void MX_GPIO_Init(void);

/* USER CODE BEGIN PFP */
// Private function prototypes

/* USER CODE END PFP */

/* USER CODE BEGIN 0 */

/* USER CODE END 0 */

int main(void)
{

  /* USER CODE BEGIN 1 */

  /* USER CODE END 1 */

  // MCU Configuration

  // Reset of all peripherals, Initializes the Flash interface and the
  // Systick.
  HAL_Init();

  /* USER CODE BEGIN Init */

  /* USER CODE END Init */

  /* Configure the system clock */
  SystemClock_Config();

  /* USER CODE BEGIN SysInit */

  /* USER CODE END SysInit */

  /* Initialize all configured peripherals */
  MX_GPIO_Init();

  /* USER CODE BEGIN 2 */

  /* USER CODE END 2 */
```

```
  /* Infinite loop */
  /* USER CODE BEGIN WHILE */
  while (1)
  {
  /* USER CODE END WHILE */

  /* USER CODE BEGIN 3 */

  }
  /* USER CODE END 3 */
}

// System Clock Configuration
void SystemClock_Config(void)
{
  RCC_OscInitTypeDef RCC_OscInitStruct;
  RCC_ClkInitTypeDef RCC_ClkInitStruct;

  // Initialize the CPU, AHB and APB buss clocks
  RCC_OscInitStruct.OscillatorType = RCC_OSCILLATORTYPE_HSI;
  RCC_OscInitStruct.HSIState = RCC_HSI_ON;
  RCC_OscInitStruct.HSICalibrationValue = 16;
  RCC_OscInitStruct.PLL.PLLState = RCC_PLL_ON;
  RCC_OscInitStruct.PLL.PLLSource = RCC_PLLSOURCE_HSI;
  RCC_OscInitStruct.PLL.PLLMUL = RCC_PLL_MUL16;
  if (HAL_RCC_OscConfig(&RCC_OscInitStruct) != HAL_OK)
  {
     _Error_Handler(__FILE__, __LINE__);
  }

  // Initialize the CPU, AHB and APB buss clocks
  RCC_ClkInitStruct.ClockType =
  RCC_CLOCKTYPE_HCLK|RCC_CLOCKTYPE_SYSCLK
  |RCC_CLOCKTYPE_PCLK1|RCC_CLOCKTYPE_PCLK2;
  RCC_ClkInitStruct.SYSCLKSource = RCC_SYSCLKSOURCE_PLLCLK;
  RCC_ClkInitStruct.AHBCLKDivider = RCC_SYSCLK_DIV1;
  RCC_ClkInitStruct.APB1CLKDivider = RCC_HCLK_DIV2;
  RCC_ClkInitStruct.APB2CLKDivider = RCC_HCLK_DIV1;

  if (HAL_RCC_ClockConfig(&RCC_ClkInitStruct, FLASH_LATENCY_2) !=
  HAL_OK)
  {
     _Error_Handler(__FILE__, __LINE__);
  }

  // Configure the Systick interrupt time
  HAL_SYSTICK_Config(HAL_RCC_GetHCLKFreq()/1000);

  // Configure the Systick
  HAL_SYSTICK_CLKSourceConfig(SYSTICK_CLKSOURCE_HCLK);
```

```
  /* SysTick_IRQn interrupt configuration */
  HAL_NVIC_SetPriority(SysTick_IRQn, 0, 0);
}

/** Configure pins as
          * Analog
          * Input
          * Output
          * EVENT_OUT
          * EXTI
      PA2    ------> USART2_TX
      PA3    ------> USART2_RX
*/

static void MX_GPIO_Init(void)
{
  GPIO_InitTypeDef GPIO_InitStruct;

  /* GPIO Ports Clock Enable */
  __HAL_RCC_GPIOC_CLK_ENABLE();
  __HAL_RCC_GPIOF_CLK_ENABLE();
  __HAL_RCC_GPIOA_CLK_ENABLE();
  __HAL_RCC_GPIOB_CLK_ENABLE();

  /*Configure GPIO pin Output Level */
  HAL_GPIO_WritePin(LD2_GPIO_Port, LD2_Pin, GPIO_PIN_RESET);

  /*Configure GPIO pin : B1_Pin */
  GPIO_InitStruct.Pin = B1_Pin;
  GPIO_InitStruct.Mode = GPIO_MODE_IT_FALLING;
  GPIO_InitStruct.Pull = GPIO_NOPULL;
  HAL_GPIO_Init(B1_GPIO_Port, &GPIO_InitStruct);

  /*Configure GPIO pins : USART_TX_Pin USART_RX_Pin */
  GPIO_InitStruct.Pin = USART_TX_Pin|USART_RX_Pin;
  GPIO_InitStruct.Mode = GPIO_MODE_AF_PP;
  GPIO_InitStruct.Pull = GPIO_NOPULL;
  GPIO_InitStruct.Speed = GPIO_SPEED_FREQ_LOW;
  GPIO_InitStruct.Alternate = GPIO_AF7_USART2;
  HAL_GPIO_Init(GPIOA, &GPIO_InitStruct);

  /*Configure GPIO pin : LD2_Pin */
  GPIO_InitStruct.Pin = LD2_Pin;
  GPIO_InitStruct.Mode = GPIO_MODE_OUTPUT_PP;
  GPIO_InitStruct.Pull = GPIO_NOPULL;
  GPIO_InitStruct.Speed = GPIO_SPEED_FREQ_LOW;
  HAL_GPIO_Init(LD2_GPIO_Port, &GPIO_InitStruct);

}

/* USER CODE BEGIN 4 */
```

```
/* USER CODE END 4 */

/**
  * @brief   This function is executed in case of error occurrence.
  * @param   None
  * @retval None
  */
void _Error_Handler(char * file, int line)
{
  /* USER CODE BEGIN Error_Handler_Debug */
  /* User can add his own implementation to report the HAL error
return state */
  while(1)
  {
  }
  /* USER CODE END Error_Handler_Debug */
}

#ifdef USE_FULL_ASSERT

/**
  * @brief Reports the name of the source file and the source line
number
  * where the assert_param error has occurred.
  * @param file: pointer to the source file name
  * @param line: assert_param error line source number
  * @retval None
  */
void assert_failed(uint8_t* file, uint32_t line)
{
  /* USER CODE BEGIN 6 */
  /* User can add his own implementation to report the file name and
line number,
     ex: printf("Wrong parameters value: file %s on line %d\r\n", file,
line) */
  /* USER CODE END 6 */

}

#endif

/**
  * @}
  */

/**
  * @}
  */

/**** (C) COPYRIGHT STMicroelectronics *****END OF FILE****/
```

The main method, as I have previously mentioned, contains most of the functional code required by the program. The following code snippet in the main method is what I refer to as a forever loop:

```
while (1)
  {
  /* USER CODE END WHILE */
  /* USER CODE BEGIN 3 */
  }
  /* USER CODE END 3 */
```

The while (1) statement causes everything contained within its scope braces to be repeated indefinitely or as I call it "forever." This is a necessary software construct because the majority of embedded systems do not use an operating system (OS) and any running program would simply run once and then halt and likely crash. Functional code, such as for blinking an LED, would be placed in this while loop. The automatic comments generated by the CubeMX application will help guide you in placing your own code in the proper spots.

There are also a series of method calls in main that both initialize and configure the internal peripherals and GPIO pins for the MCU. These are listed in the same sequence as they are called:

- HAL_Init();—Initialize the Hardware Abstraction Layer (HAL) framework
- SystemClock_Config();—Initialize and configure the clock system
- MX_GPIO_Init();—Initialize and configure all enabled GPIO pins

There can be additional methods added to the main.c file depending upon the number and types of internal peripherals that are enabled. In addition, some extra code will have to be added to files outside of main.c to support operations such as interrupts. I will cover those cases in later chapters.

At this point, I need to discuss the ARM CMSIS software architecture in order for you to understand how the newly·generated code fits in with the overall software architecture.

ARM Cortex Microcontroller Software Interface Standard (CMSIS)

ARM, which holds the intellectual property (IP) rights to all the Cortex-based MCUs, has created an interface standard that allows for efficient cross-vendor software development. This standard is called the Cortex Microcontroller Software Interface Standard and usually abbreviated as CMSIS. CMSIS addresses a full set of development tools including compilers, run-time libraries, debuggers, and other associated tools and utilities. CMSIS consists of the following main components:

- CMSIS-CORE: Standard application programming interface (API) covering the Cortex M0, M3, M4, and M7 processors and associated peripherals.

- CMSIS-Driver: Peripheral driver software interfacing all applicable devices with supported middleware software. This CMSIS portion contains items such as the file systems, communication stacks, and user interfaces.

- CMSIS-DSP: Extensive optimized library supporting different data including both fixed and floating point data types. Library is applicable to Cortex M0, M3, M4, and M7 core processors.

- CMSIS-RTOS: A standardized API implementing a real-time operating system (RTOS). This API provides a framework that supports a variety of specific Cortex, RTOS implementations.

- CMSIS-DAP: Debug access port (DAP) is a standardized set of firmware routines that implements debugging features optimized for Cortex processors. It connects to the CoreSight Debug Access Port to have rapid access to real-time data.

- CMSIS-SVD: This is an extensible markup language (XML) file named System View Description (SVD), which contains peripheral descriptions suitable for use by programming language header (include) files as well as debuggers and interrupt controllers.

- CMSIS-Pack: An XML file collection containing logical descriptions and parameters for a variety of devices. A main file named PDSC is used to extract desired data from the XML file collection.

It should be noted that the STM hardware abstraction layer (HAL), discussed in the next chapter, contains specifics only related to STM MCUs and is somewhat noncompliant with the ARM Cortex Microcontroller Software Interface Standard (CMSIS). Using the CMSIS is totally voluntary on part of vendors manufacturing ARM-based MCUs and STM has apparently decided that using HAL is the best approach in its case to handle the wide variety of MCUs and associated development boards it makes. I believe STM tried to stay compliant with the CMSIS but still needed to support its vast array of products.

CubeMX-Generated C Code

It is now time to return to a discussion of the automatically generated code that the CMSIS framework has introduced. Figure 3-24 is an interesting composite of the newly generated directories that show how they are related to the various CMSIS components.

The Application/User directory contains essentially skeleton files that present a framework that users can populate with specific code to implement a desired application. The files contained within these directories are only applicable to the Application block, as shown in Figure 3-24.

The Drivers/CMSIS structure contains header files with names such as core_cm3, core_cm4, and so on, which contain specific logical definitions required by the real MCU peripherals and all associated control/data registers. There are also core MCU definitions as well as interrupt controller definitions and helper functions.

The Drivers/STM32F3xx_Hal_Driver directory contains all the remaining data and parameters necessary for MCU operations not already provided in other directory files. This data covers items such as interrupt numbers, peripheral device addresses, restart vectors, and additional helper functions.

The Application/MDK-ARM contains ARM-specific startup code necessary to initialize the target MCU.

Compiling and Downloading the Project

At this point there is sufficient initialization and configuration code to build a downloadable file suitable to be run on the target. Unfortunately, it is a null

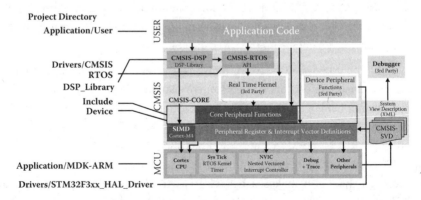

Figure 3-24   *Generated code related to CMSIS structure.*

project, meaning that nothing meaningful is being accomplished by the MCU. However, it is still a useful exercise to determine if the code, as it stands, can be compiled and downloaded to the target.

Figure 3-25 shows where the Keil IDE build icon is located. Clicking on this icon will start the automatic build process converting the source code in the project to a binary file.

Figure 3-26 shows all the intermediate results that happen during the build process. This log can be very useful in the case of errors that can happen during the build process.

Figure 3-25   *Keil IDE build icon.*

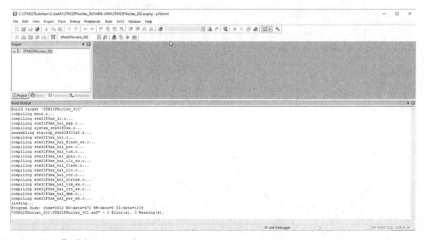

Figure 3-26   *Build process log.*

You may have noticed that the final compiled and linked binary file is in an axf format. This format is not compatible to be used with the ST-LINK, which expects the download file to have a hex format. This is easily fixed by clicking on the Flash menu selection and then clicking on the Configure Flash Tools. Click on the Output tab once the flash tools selection has opened. Then click on the checkbox next to the Create HEX File selection. Figure 3-27 shows this selection.

The project must be rebuilt once the hex file has been selected. This is easily done by just again clicking on the build icon. The IDE is smart enough to understand that the source code does not have to be recompiled but simply linked again to form a hex file. The resulting hex file is clearly shown in Figure 3-28.

The target can now be programmed once the hex file is created. The ST-LINK application is used to download the hex file into the target, as has been previously described in Chapter 2.

Downloading the Hex Code

You first need to open the ST-LINK application and open the appropriate hex file to be downloaded. The hex file is automatically placed into the project's MDK-ARM subdirectory as part of the build process. You will have to browse to this

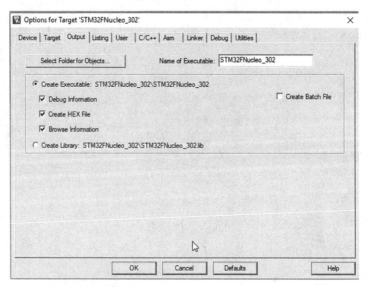

Figure 3-27 *Select hex file output.*

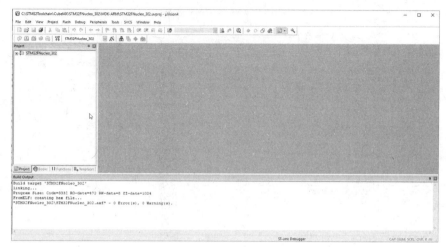

Figure 3-28   *Hex file output.*

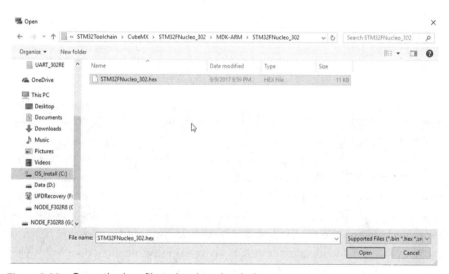

Figure 3-29   *Open the hex file to be downloaded.*

directory using the ST-LINK open file feature found in the File menu selection. Figure 3-29 shows how the sample hex file was selected.

The target must then be connected to the host computer running the ST-LINK application using an appropriate USB cable, as has been previously discussed. You will next need to click on the Target menu selection and then click on the Connect

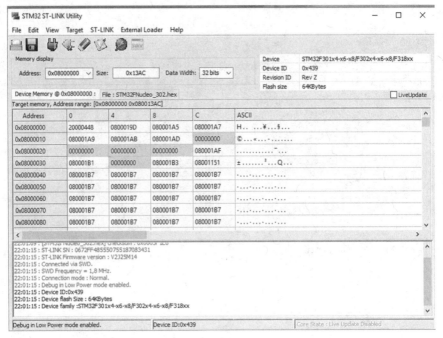

Figure 3-30 *ST-LINK connected to a target.*

selection. Figure 3-30 shows the ST-LINK window after the application is connected to the target.

All that's left is to download the hex file that has been loaded into the ST-LINK application. Click on the Target menu selection and then click on the Program and Verify selection. Figure 3-31 shows the Program and Verify dialog box.

Click on the Start button to download the hex file into the target board. The red/green LED on the Nucleo board will start flashing with alternating colors to indicate that the download operation is progressing. While the download is very quick, the LED will continue to flash. I guess the STM engineers just wanted to let users know the board was programmed. Figure 3-32 shows the result of a successful download.

The target board should be functioning in accordance with its program. In this case, absolutely nothing will happen because no functionality was programmed. This addition of functionality to the program will be investigated in the next chapter. However, I want to ensure that the download was okay, so I previously programmed the target to flash the user LED. Once the new hex file was downloaded,

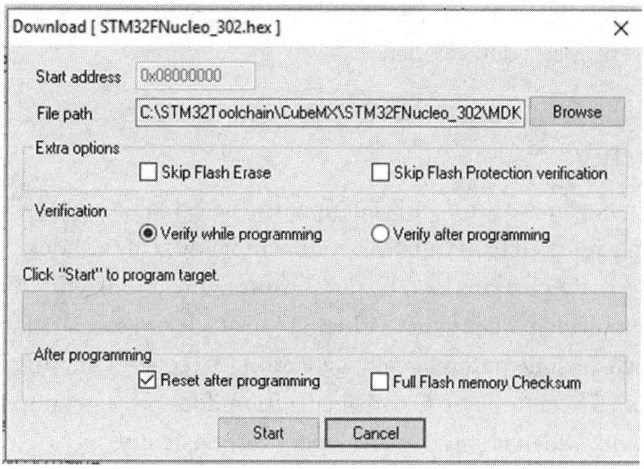

Figure 3-31 *Program and Verify dialog box.*

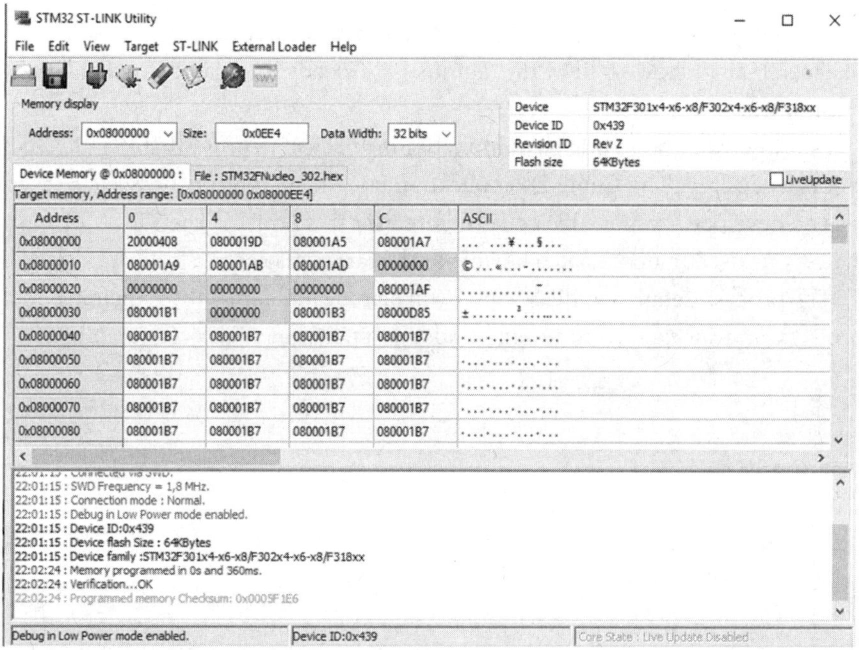

Address	0	4	8	C	ASCII
0x08000000	20000408	0800019D	080001A5	080001A7	... ...¥...§...
0x08000010	080001A9	080001AB	080001AD	00000000	©...«...-.......
0x08000020	00000000	00000000	00000000	080001AF	............¯...
0x08000030	080001B1	00000000	080001B3	08000D85	±.......³.....…
0x08000040	080001B7	080001B7	080001B7	080001B7	·...·...·...·...
0x08000050	080001B7	080001B7	080001B7	080001B7	·...·...·...·...
0x08000060	080001B7	080001B7	080001B7	080001B7	·...·...·...·...
0x08000070	080001B7	080001B7	080001B7	080001B7	·...·...·...·...
0x08000080	080001B7	080001B7	080001B7	080001B7	·...·...·...·...

Figure 3-32 *Successful hex file download.*

the LED stopped blinking, which indirectly confirmed that the nonfunctioning program was properly downloaded.

Summary

The chapter began with an introduction to the STM32CubeMX application, which I simply referred to as CubeMX for the remainder of the book. The application has four main portions or tabs that I thoroughly discussed. CubeMX's purpose is to initialize and configure either STM MCUs or associated development boards, which include automatically generating C code to support the desired configuration. The automatic C code generation does not extend to any embedded application, which is still the developer's responsibility.

A generic CubeMX project was created that did not do anything other than to provide a platform for the automatically generated C code, which supports the peripherals, clocks, communication channels, and interrupts for the specific Nucleo development board. I also presented the code contained in the main.c file and briefly discussed the contents and their operations.

I next discussed the ARM CMSIS, which most ARM vendors try to be in compliance. I also showed how the automatic C code is related to the CMSIS framework.

I showed you how to build a nonfunctioning project to prove that the toolchain works as desired. The CubeMX application created all the C code because there was no developer-responsible code required for this project. The ST-LINK application was used to download a hex file to the target platform.

The process detailed in this chapter will be repeated for all the remaining book projects, and specialized code will be added to perform all desired functions.

4

STM Project Development

In this chapter I will demonstrate how to build C/C++ language projects with a STM Nucleo board using the toolchain created in Chapter 2 and the CubeMX application described in Chapter 3. I will be using C/C++ language to create the source code for these projects. It would be helpful, but not really required, if readers are somewhat familiar with the C/C++ language in order to achieve a better understanding of the entire process. I will go thoroughly into the example code for the "Hello World" project to help clarify what is happening with the source code for those readers who are not too proficient in using C.

Hello World Project

The traditional and customary "Hello World" program is the one that beginning developers almost always make when starting their computer language studies. In the embedded world, this program has morphed into blinking an LED instead of displaying "Hello World!" on a monitor screen. I will demonstrate step-by-step how to create a "Hello World" project using a Nucleo64-STMF302R8 demonstration board connected to a Windows 10-based host system running the toolchain established in Chapter 2. You will not benefit much from this chapter unless you have read the previous chapters and created a working toolchain and become somewhat comfortable using the CubeMX application.

Creating the Hello Nucleo Project

The first step in creating a project is to open CubeMX application. Once the application is started, click on New Project and click on the Board tab. Select the

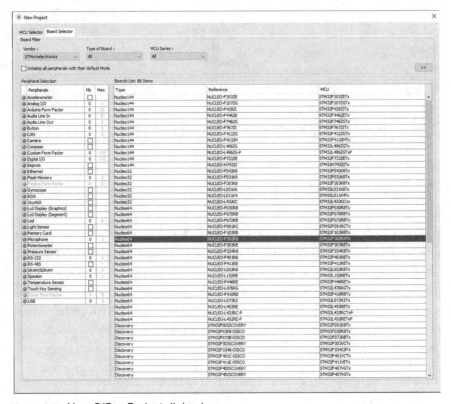

Figure 4-1   *New C/C++ Project dialog box.*

Nucleo board that you are using. In my case, it is the NUCLEO-F302R8 board, as shown in Figure 4-1.

Next, double-clicking on the board selection line will cause the project pinout screen shown in Figure 4-2 to be displayed.

The next step is to click on the Project menu item, which will result in the dropdown menu, as shown in Figure 4-3.

Clicking on the Generate Code selection will cause the Project Settings dialog box appear, as shown in Figure 4-4.

You need to provide a name in the Project Name textbox. It should be something relevant to the intended project purpose. Also confirm that the Project Location is correct along with the Toolchain Folder Location.

It is very important that you select the MDK-ARM V4 from the Toolchain/IDE combo list. The project cannot be created without MDK-ARM V4 selection.

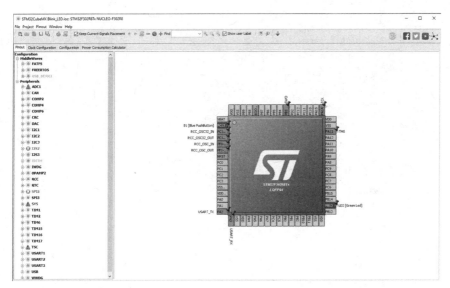

Figure 4-2   *Project pinout screen.*

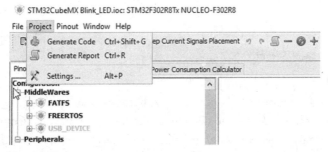

Figure 4-3   *Project dropdown menu.*

The default Linker Settings should be okay and the "Mcu and Firmware Package" and "Firmware Package Name and Version" should also be similar to what is shown in the figure.

Finally, ensure that the checkbox next to "Use Default Firmware Location" is checked, otherwise the IDE will not be able to create the binary file.

Do NOT click on the Ok button, but instead click on the Code Generator tab near the top of the dialog box. This should make the Project Settings/Code Generator dialog box appear, as shown in Figure 4-5.

Figure 4-4   *Project Settings dialog box.*

The radio button next to "Copy all used libraries into the project folder" will initially be selected. You should click on the radio button next to "Copy only the necessary library files" in order to minimize the size of the overall project. Neglecting to make this change will not affect the size of the final binary file, but will needlessly enlarge the project folder size.

Now you can click on the Ok button to have the CubeMX application automatically generate the C/C++ code.

After a short interval, a successful Code Generation dialog box should appear, as shown in Figure 4-6.

Clicking on the Open Project button will cause the Keil IDE to be opened with the newly created project shown in the Project pane to the left side of the window, as shown in Figure 4-7.

The error dialog that appears in Figure 4-7 has been discussed in the previous chapter and is due to the IDE not having a reference to the real target hardware.

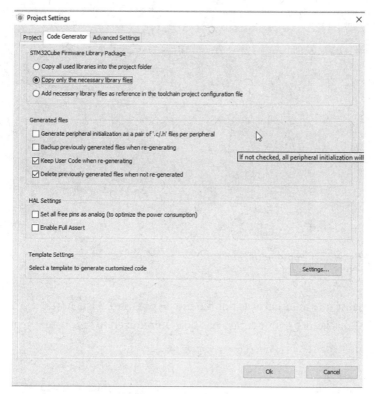

Figure 4-5 *Project Settings/Code Generator dialog box.*

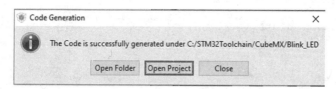

Figure 4-6 *Successful Code Generation dialog box.*

That is easily fixed by going to the Device Database, as described in Chapter 3. Once that is resolved, you should expand the Project directory by clicking on the + check box. Once expanded, click on the dropdown arrow next to the Application/ User subdirectory to reveal the files in that directory. Finally, double-click on the main.c file, which will display the following listing. Note that this listing is the same, as has been previously shown in Chapter 3. However, I do this to provide the proper context to make the desired modifications in order to add some

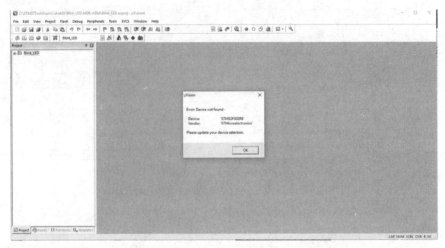

Figure 4-7 *Keil IDE with the new project.*

functionality into an otherwise nonfunctional program. Also, note that the extensive STM disclaimer commentary has been minimized to save listing space.

```
[1]/** The STM disclaimer goes here */

// Includes
[2]#include "main.h"
[3]#include "stm32f3xx_hal.h"

/* USER CODE BEGIN Includes */

/* USER CODE END Includes */

// Private variables

/* USER CODE BEGIN PV */
// Private variables

/* USER CODE END PV */

// Private function prototypes
[4]void SystemClock_Config(void);
static void MX_GPIO_Init(void);

/* USER CODE BEGIN PFP */
// Private function prototypes

/* USER CODE END PFP */
```

```
/* USER CODE BEGIN 0 */

/* USER CODE END 0 */

[5]int main(void)
[6]{

  /* USER CODE BEGIN 1 */

  /* USER CODE END 1 */

  // MCU Configuration

  // Reset of all peripherals, Initializes the Flash interface and the
  // Systick.
  HAL_Init();

  /* USER CODE BEGIN Init */

  /* USER CODE END Init */

  /* Configure the system clock */
  [7]SystemClock_Config();

  /* USER CODE BEGIN SysInit */

  /* USER CODE END SysInit */

  /* Initialize all configured peripherals */
  [8]MX_GPIO_Init();

  /* USER CODE BEGIN 2 */

  /* USER CODE END 2 */

  /* Infinite loop */
  /* USER CODE BEGIN WHILE */
  [9]while (1)
  {
  /* USER CODE END WHILE */

  /* USER CODE BEGIN 3 */

  }
  /* USER CODE END 3 */
}

// System Clock Configuration
void SystemClock_Config(void)
{
```

```
RCC_OscInitTypeDef RCC_OscInitStruct;
RCC_ClkInitTypeDef RCC_ClkInitStruct;

// Initialize the CPU, AHB and APB buss clocks
RCC_OscInitStruct.OscillatorType = RCC_OSCILLATORTYPE_HSI;
RCC_OscInitStruct.HSIState = RCC_HSI_ON;
RCC_OscInitStruct.HSICalibrationValue = 16;
RCC_OscInitStruct.PLL.PLLState = RCC_PLL_ON;
RCC_OscInitStruct.PLL.PLLSource = RCC_PLLSOURCE_HSI;
RCC_OscInitStruct.PLL.PLLMUL = RCC_PLL_MUL16;
if (HAL_RCC_OscConfig(&RCC_OscInitStruct) != HAL_OK)
{
  _Error_Handler(__FILE__, __LINE__);
}

// Initialize the CPU, AHB and APB buss clocks
RCC_ClkInitStruct.ClockType =
RCC_CLOCKTYPE_HCLK|RCC_CLOCKTYPE_SYSCLK
|RCC_CLOCKTYPE_PCLK1|RCC_CLOCKTYPE_PCLK2;
RCC_ClkInitStruct.SYSCLKSource = RCC_SYSCLKSOURCE_PLLCLK;
RCC_ClkInitStruct.AHBCLKDivider = RCC_SYSCLK_DIV1;
RCC_ClkInitStruct.APB1CLKDivider = RCC_HCLK_DIV2;
RCC_ClkInitStruct.APB2CLKDivider = RCC_HCLK_DIV1;

if (HAL_RCC_ClockConfig(&RCC_ClkInitStruct, FLASH_LATENCY_2) !=
HAL_OK)
{
  _Error_Handler(__FILE__, __LINE__);
}

// Configure the Systick interrupt time
HAL_SYSTICK_Config(HAL_RCC_GetHCLKFreq()/1000);

// Configure the Systick
[7]HAL_SYSTICK_CLKSourceConfig(SYSTICK_CLKSOURCE_HCLK);

/* SysTick_IRQn interrupt configuration */
HAL_NVIC_SetPriority(SysTick_IRQn, 0, 0);
}

/** Configure pins as
        * Analog
        * Input
        * Output
        * EVENT_OUT
        * EXTI
    PA2   ------> USART2_TX
    PA3   ------> USART2_RX
*/

static void MX_GPIO_Init(void)
```

```
{
  GPIO_InitTypeDef GPIO_InitStruct;

  /* GPIO Ports Clock Enable */
  __HAL_RCC_GPIOC_CLK_ENABLE();
  __HAL_RCC_GPIOF_CLK_ENABLE();
  __HAL_RCC_GPIOA_CLK_ENABLE();
  __HAL_RCC_GPIOB_CLK_ENABLE();

  /*Configure GPIO pin Output Level */
  HAL_GPIO_WritePin(LD2_GPIO_Port, LD2_Pin, GPIO_PIN_RESET);

  /*Configure GPIO pin : B1_Pin */
  GPIO_InitStruct.Pin = B1_Pin;
  GPIO_InitStruct.Mode = GPIO_MODE_IT_FALLING;
  GPIO_InitStruct.Pull = GPIO_NOPULL;
  HAL_GPIO_Init(B1_GPIO_Port, &GPIO_InitStruct);

  /*Configure GPIO pins : USART_TX_Pin USART_RX_Pin */
  GPIO_InitStruct.Pin = USART_TX_Pin|USART_RX_Pin;
  GPIO_InitStruct.Mode = GPIO_MODE_AF_PP;
  GPIO_InitStruct.Pull = GPIO_NOPULL;
  GPIO_InitStruct.Speed = GPIO_SPEED_FREQ_LOW;
  GPIO_InitStruct.Alternate = GPIO_AF7_USART2;
  HAL_GPIO_Init(GPIOA, &GPIO_InitStruct);

  /*Configure GPIO pin : LD2_Pin */
  GPIO_InitStruct.Pin = LD2_Pin;
  GPIO_InitStruct.Mode = GPIO_MODE_OUTPUT_PP;
  GPIO_InitStruct.Pull = GPIO_NOPULL;
  GPIO_InitStruct.Speed = GPIO_SPEED_FREQ_LOW;
  HAL_GPIO_Init(LD2_GPIO_Port, &GPIO_InitStruct);

}

/* USER CODE BEGIN 4 */

/* USER CODE END 4 */

/**
  * @brief  This function is executed in case of error occurrence.
  * @param  None
  * @retval None
  */
void _Error_Handler(char * file, int line)
{
  /* USER CODE BEGIN Error_Handler_Debug */
  /* User can add his own implementation to report the HAL error re-
turn state */
  while(1)
  {
```

```
    }
    /* USER CODE END Error_Handler_Debug */
}

#ifdef USE_FULL_ASSERT

/**
    * @brief Reports the name of the source file and the source line
number
    * where the assert_param error has occurred.
    * @param file: pointer to the source file name
    * @param line: assert_param error line source number
    * @retval None
    */
void assert_failed(uint8_t* file, uint32_t line)
{
    /* USER CODE BEGIN 6 */
    /* User can add his own implementation to report the file name and
line number,
        ex: printf("Wrong parameters value: file %s on line %d\r\n", file,
line) */
    /* USER CODE END 6 */

}

#endif
/**
    * @}
    */

/**
    * @}
    */


/**** (C) COPYRIGHT STMicroelectronics *****END OF FILE****/
```

I have chosen to provide commentary to the listing using italicized number annotations to designate those portions of the listing being discussed. Simply refer to the listing and the associated number in italics to keep track of the discussion topic.

[1]—Either the /* ... */ or // designate a commented line, which is ignored by the compiler. The first format allows for multiple comment lines, while the second is only for a single line. Comments are for human use as a means for the developer to explain what is happening within the source code. I strongly recommend that you always comment your programs as you see with this listing.

[2]—The # symbol indicates a compiler instruction or directive immediately follows. In this case the statement is #include "main.h" which instructs

the compiler to use a header file named main.h. This particular header file is stored in a workspace header file directory because the header file name is enclosed by double quotes. The main.h contains constants, definitions, and functions that provide "understandable" names to general-purpose input/output (GPIO) pins such as LD2 for the user LED and Push Button for the user push button.

[3]—This header file provides all the required constants and definitions for the project code to utilize the HAL framework.

[4]—All methods that are called must be declared prior to use in the C/C++ language. That is the purpose of the prototype section.

[5]—This is the starting point for the program. The int in front of the main function name indicates that the main function returns an integer value. This returned value to the operating system is normally a 0 if the program terminated normally, otherwise a 1 is typically returned.

The function arguments

```
(int argc, char* argv[])
```

are used only when starting values are to be supplied to the program before it starts to execute. In this case, they are not required.

[6]—This is the opening brace. All C/C++ functions enclose all their code between opening and closing braces. There must be a matching closing brace for this main function or else the C/C++ compiler will generate an error.

[7]—This statement starts a system timer that generates a SysTick every 1 ms. Note the statement ends with a semicolon, which is normally required for all C/C++ declarative statements. Omitting semicolons is a common mistake done by beginning (and not so beginning) developers.

[8]—This statement initializes the appropriate GPIO pin that directly controls the user LED. In this case, the GPIO pin is PA5, meaning it is one of the eight pins found in port A. There are multiple GPIO ports in STM MCUs. The exact amount depends on the particular model.

[9]—This is the start of an infinite loop, meaning the program will run forever until manually stopped or having the power disconnected. The 1 in the parenthesis indicates a true value. Often, developers will use a Boolean variable instead of a fixed constant, thus allowing the program to be stopped programmatically from within the scope of the while construct. This scope is determined by opening and closing braces, similar to what was used with the main function. This use of multiple brace pairs is known as nesting.

I will provide additional explanations regarding the code modifications.

I hope that this brief review of the main.c source code has provided you a little insight into how a C/C++ program functions.

Adding Functionality to the Program

I will show you how to add a very simple function to this program, which will flash the onboard green user LED at a rate of once per second. The two statements to be added are shown in the following code snippet, which is the forever loop.

```
while (1)
  {
  /* USER CODE END WHILE */

  /* USER CODE BEGIN 3 */
  HAL_GPIO_TogglePin(LD2_GPIO_Port, LD2_Pin);
  HAL_Delay(500);

  }
  /* USER CODE END 3 */
```

The first statement HAL_GPIO_TogglePin(LD2_GPIO_Port, LD2_Pin); toggles the GPIO pin that is permanently connected to the user LED on the selected Nucleo board. Note the use of human recognizable names for both the GPIO pin connected to the LED and the associated GPIO port. That feature is made possible by the defines contained in the main.h file, which is included in the main.c file.

The second statement HAL_Delay(500); pauses the processor for 500 ms before the LED state changes. This will cause the LED to blink once per second with a 50% duty cycle, meaning it is on for 0.5 s and off for 0.5 s.

Compiling and Executing the Modified Program

Just click on the build icon after you added the additional two statements in the forever loop. Also, ensure you have enabled the hex file output, as described in the previous chapter. This is required prior to using the ST-LINK utility.

Again, follow the ST-LINK detailed instructions provided in Chapter 3 on how to load the hex file and download it into the target board.

The user LED should immediately start to blink once the hex file is downloaded. I am fairly confident that you should have no problem with this procedure given that the program built successfully.

Simple Modification for the main.c Function

The simplest modification to the main.c function is to change the LED timing duration, which is what I have done for this next demonstration. The following code snippet shows the modifications made to the on and off LED timing duration. It was changed to 2 s:

```
while (1)
  {
  /* USER CODE END WHILE */

  /* USER CODE BEGIN 3 */
  HAL_GPIO_TogglePin(LD2_GPIO_Port, LD2_Pin);
  HAL_Delay(2000);

  }
  /* USER CODE END 3 */
```

The source code in the main.c file was first changed and then saved. I then rebuilt the project, which in turn generated a new hex file. This hex new file was then loaded into the ST-LINK utility and subsequently downloaded into the Nucleo board using the process I previously described. I immediately observed that the user LED was then flashing at a rate of 2 s on and 2 s off after the new hex file was downloaded into the target.

Complex Modification for the main.c File

The next demonstration is a bit more complex than the previous one in that I want to show you how to blink the LED with different on and off times. I will still be using the user LED LD2 as the light source but will modify the main.c file to incorporate two different timing intervals.

I choose to use several define statements to create different on and off durations for the LED. The main function also uses an if/else statement that causes the on and off intervals to alternate based on the truth value of a Boolean variable named sw. The following is the code snippet for the modified portion of the main.c file:

```
// ----- Timing definitions ---------------------------------------
#define BLINK_TICKS1  (500)
#define BLINK_TICKS2  (1500)

int sw =1;
  // Infinite loop
```

```
while (1)
  {
  if(sw) {
      HAL_GPIO_TogglePin(LD2_GPIO_Port, LD2_Pin);
      HAL_Delay(BLINK_TICKS1);
      sw = 0;
  }
  else {
      HAL_GPIO_TogglePin(LD2_GPIO_Port, LD2_Pin);
      HAL_Delay(BLINK_TICKS2);
      sw = 1;
  }
}
```

After the main.c file was modified, the project was built in the manner previously described and the resultant hex file was downloaded into the Nucleo board using the ST-LINK utility. The user LED immediately started blinking on for 0.5 s and off for 1.5 s for a net 2-s blink rate. The on and off durations may easily be modified by changing the appropriate define statements.

This last demonstration project concludes this initial discussion on how to create and modify embedded software projects for a STM Nucleo-64 board. The next chapter explores how to use additional GPIOs and the use of interrupts, which will greatly expand the capabilities of the Nucleo board.

Summary

This chapter contains three demonstration C/C++ language projects that all blinked an on-board LED installed on a Nucleo64-STM32F302R8 board. Each project progressed in complexity from an extremely simple constant rate LED flashing to a more complex multi-interval flashing of the LED.

I explained using a step-by-step approach how to conduct a project build. The final hex file created by the build process was downloaded into the Nucleo-64 board using the ST-LINK utility.

A brief discussion of how a C/C++ main function works was also presented for readers who might be a bit "rusty" on their C/C++ language prowess.

5

General-Purpose Input Output (GPIO) and the STM Hardware Abstraction Layer (HAL)

As the formidable title states, this chapter deals with how GPIO operates on a STM Nucleo-64 board and how the HAL software is involved. It is important to first study the STM MCU architecture in order to fully understand how the GPIO ports function and their relationship to HAL. I would also acknowledge that many of the figures and table content are sourced from either the STM32F302R8 datasheet or reference manual. I would highly recommend that you download and have them readily available to help fill in any gaps in my discussions.

Figure 5-1 is the STM32F302R8 block diagram, which is the MCU used in the Nucleo-64 development board used in this book.

Note that there are six banks of GPIO ports shown on the left side of Figure 5-1, which are connected to the bidirectional advanced high-speed bus 1 (AHB1). This bus transfers digital data internally at a rate up to a maximum of 100 Mbps. HAL effectively manages the data transfers from the GPIO ports as well as from the many other peripheral components depicted in the figure.

I will start the HAL discussion by examining perhaps the simplest HAL module, which is the HAL_GPIO. This module has already been used multiple times

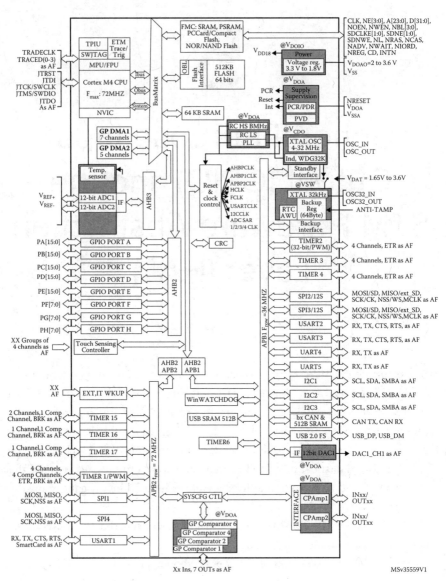

Figure 5-1   *STM32F302R8 block diagram.*

in programs discussed in earlier chapters. The HAL_GPIO module was used without explanation simply because it was necessary to implement a required function such as blinking an LED or reading the state of a push-button switch. I will shortly closely examine the HAL_GPIO module, but first I will discuss the concept of memory-mapped peripherals.

Memory-Mapped Peripherals

STM MCU peripherals are controlled and data is transferred using addresses within a 32-bit range. This is known as memory mapping because it makes no difference from a programming perspective whether it is communicating with a memory location or a peripheral register. The 32-bit address range is usually represented by eight integer digits, meaning the total possible range of locations is 0x0000 0000 to 0xFFFF FFFF. Note that the hex number is preceded by a 0x and a space is paced after four of the leading hex digits. The 0x is a common programming construct indicating that the number is in a hex format and the space between the two hex number groups is just to make the number more readable. This address format is also used throughout the STM documentation.

Figure 5-2 shows the major blocks that constitute the entire 32-bit address range.

Also shown in Figure 5-2 are two breakout blocks, which detail the core memory locations as well as the peripheral/bus locations. I will next briefly discuss the core memory locations, which are important in their own right, but not very relevant to the peripheral address locations.

Core Memory Addresses

Table 5-1 shows the types of memory and their preassigned memory addresses.

One question that many readers might be pondering is how a PC can handle 8, 16, or even 32 GB of SRAM with only 32 bits of addressing capability? The answer lies in using both an operating system designed for extended memory accesses and a hardware unit known as a memory management unit (MMU). All the major PC OSs such as Windows, OSX, and Linux have extended memory functionality built-in that allows for 32-bit addressing to extend into multiple blocks or pages. You could think of using a 4-bit register, where each bit would activate a particular block or page. Therefore, bit 0 would allow access to all the base addresses starting

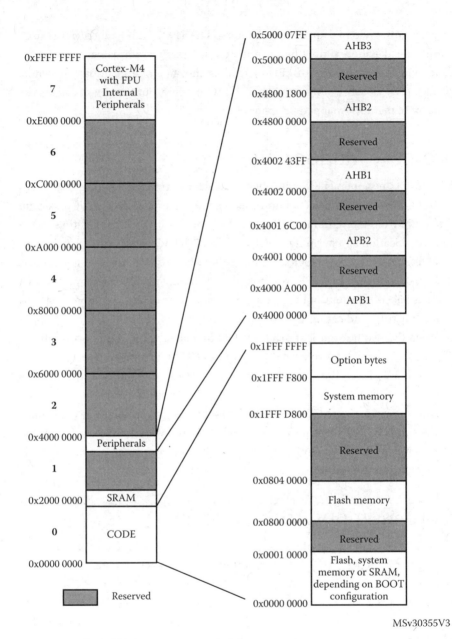

Figure 5-2 *Complete 32-bit STM address range.*

Memory Type	Address Range	Size (KB)	Remark
Flash	0x0800 0000 to 0x0803 FFFF	64	–
System	0x1FFF D800 to 0x1FFF E7FF	24	Used with an operating system
SRAM	0x2000 0000 to 0x2000 3FFF	16	–
Aliased	0x0000 0000 to 0x0000 FFFF	64	Aliased to Flash, System, or SRAM depending on the configuration of the BOOT pins

Table 5-1 *STM Core Memory Address Ranges*

at 0x0000 0000, while bit 1 would allow addressing to virtually start at 0x0001 0000 0000. Of course, there are only 32 real bits being used, but the OS handles the addresses as if they started one location beyond 0xFFFF FFFF. This 4-bit register can be thought of as being part of the MMU. Four bits means that a maximum of 64 GB can be handled in a 32-bit PC, which coincidentally is often the maximum SRAM that may be installed in such a PC. Of course, all practical physical memory limitations are relieved when a 64-bit system is used, but that requires an expensive CPU and additional energy requirements.

The preceding discussion is not applicable to a typical STM MCU because memory is quite constrained and vast amounts are not needed to handle typical embedded applications. I do want to mention that STM32F302R8 does incorporate a memory protection unit (MPU). The MPU is not a MMU, but is provided to manage CPU memory accesses such that one task will not accidentally corrupt or conflict with another task's memory accesses. Memory can be arranged in up to eight areas with each area sized between 32 B and 4 GB, the last comprising the whole addressable memory range. The MPU is normally controlled by a real-time operating system (RTOS), which has an API contained within the CMSIS, which has been introduced in the last chapter. I will not be using the MPU for any of the book projects. You should consider its use only for the most safety-critical or mission-essential applications. Think of it being used in automotive or flight safety systems.

Peripheral Memory Addresses

Figure 5-3 shows the beginning portion of the memory addresses allocated to the STM32F302R8 peripherals,

Bus	Boundary address	Size (bytes)	Peripheral
AHB3	0×5000 0000 - 0×5000 03FF	1 K	ADC1
	0×4800 1800 - 0×4FFF FFFF	~132 M	Reserved
AHB2	0×4800 1400 - 0×4800 17FF	1 K	GPIOF
	0×4800 1000 - 0×4800 13FF	1 K	Reserved
	0×4800 0C00 - 0×4800 0FFF	1 K	GPIOD
	0×4800 0800 - 0×4800 0BFF	1 K	GPIOC
	0×4800 0400 - 0×4800 07FF	1 K	GPIOB
	0×4800 0000 - 0×4800 03FF	1 K	GPIOA
	0×4002 4400 - 0×47FF FFFF	~128 M	Reserved
AHB1	0×4002 4000 - 0×4002 43FF	1 K	TSC
	0×4002 3400 - 0×4002 3FFF	3 K	Reserved
	0×4002 3000 - 0×4002 33FF	1 K	CRC
	0×4002 2400 - 0×4002 2FFF	3 K	Reserved
	0×4002 2000 - 0×4002 23FF	1 K	Flash interface
	0×4002 1400 - 0×4002 1FFF	3 K	Reserved
	0×4002 1000 - 0×4002 13FF	1 K	RCC
	0×4002 0800 - 0×4002 0FFF	2 K	Reserved
	0×4002 0400 - 0×4002 07FF	1 K	Reserved
	0×4002 0000 - 0×4002 03FF	1 K	DMA1
	0×4001 8000 - 0×4001 FFFF	32 K	Reserved
APB2	0×4001 4C00 - 0×4001 7FFF	13 K	Reserved
	0×4001 4800 - 0×4001 4BFF	1 K	TIM17
	0×4001 4400 - 0×4001 47FF	1 K	TIM16
	0×4001 4000 - 0×4001 43FF	1 K	TIM15
	0×4001 3C00 - 0×4001 3FFF	1 K	Reserved
	0×4001 3800 - 0×4001 3BFF	1 K	USART1
	0×4001 3400 - 0×4001 37FF	1 K	Reserved
	0×4001 3000 - 0×4001 33FF	1 K	Reserved
	0×4001 0800 - 0×4001 2FFF	10 K	TIM1
	0×4001 0400 - 0×4001 07FF	1 K	EXTI
	0×4001 0000 - 0×4001 03FF	1 K	SYSCFG + COMP + OPAMP
	0×4000 7C00 - 0×4000 FFFF	33 K	Reserved

Figure 5-3 *Beginning portion of the STM32F302R8 peripheral allocations.*

You should be able to see that the AHB2 bus has the address range 0x4800 0000 to 0x4FFF 17FF. What is specific to this chapter's focus is the total GPIO range of 0x4002 0000 to 0x4002 0x1FFF, which is divided into six smaller 1-KB blocks. Five of the six 1-KB blocks are allocated to GPIO ports A, B, C, D, and F, respectively. There is an unused block at range 0x4800 1000 to 0x4800 13FF, which has no significant impact on the port allocations. I have no idea why there is a reserved block allocated in what was a contiguous block allocation or why the final GPIO port is labeled F instead of E, as one might expect from the prior port designations. The port memory designation and allocation question probably can only be answered by the STM MCU chip designers. It is up to MCU users to accept the designations and allocations as they are and work with them to the best of their abilities.

I elected to include the following figure on peripheral locations for completeness sake and to provide a readily available reference for additional peripherals. Figure 5-4 shows the peripherals connected to the internal advanced peripheral bus 1 (APB1). The peripherals connected to the bus are as follows:

- Timers (TIMxx)

- Universal Synchronous/Asynchronous Receiver/Transmitters (USARTx)

- Serial peripheral interfaces (SPIx)

- Inter-integrated interfaces (I2Cx)

- Universal serial bus (USB)

- Integrated sound interface (I2S)

- Digital-to-analog converter (DAC)

- Controller area network interface (CAN)

- Real-time clock (RTC)

- Watch dog timers (IWDG, WWDG)

I will now return to the HAL_GPIO discussion after introducing to the concept of peripheral memory-mapped addresses.

HAL_GPIO Module

It would be wise to examine a sample GPIO circuit before discussing how it functions with the module software. It is always an important step to understanding the underlying hardware that implements the software.

Bus	Boundary address	Size (bytes)	Peripheral
APB1	0×4000 7800 - 0×4000 7BFF	9 K	I2C3
	0×4000 7400 - 0×4000 77FF	1 K	DAC1
	0×4000 7000 - 0×4000 73FF	1 K	PWR
	0×4000 6C00 - 0×4000 6FFF	1 K	Reserved
	0×4000 6800 - 0×4000 6BFF	1 K	Reserved
	0×4000 6400 - 0×4000 67FF	1 K	bxCAN
	0×4000 6000 - 0×4000 63FF	1 K	USB/CAN SRAM
	0×4000 5C00 - 0×4000 5FFF	1 K	USB device FS
	0×4000 5800 - 0×4000 5BFF	1 K	I2C2
	0×4000 5400 - 0×4000 57FF	1 K	I2C1
	0×4000 5000 - 0×4000 53FF	1 K	Reserved
	0×4000 4C00 - 0×4000 4FFF	1 K	Reserved
	0×4000 4800 - 0×4000 4BFF	1 K	USART3
	0×4000 4400 - 0×4000 47FF	1 K	USART2
	0×4000 4000 - 0×4000 43FF	1 K	I2S3ext
	0×4000 3C00 - 0×4000 3FFF	1 K	SPI3/I2S3
	0×4000 3800 - 0×4000 3BFF	1 K	SPI2/I2S2
	0×4000 3400 - 0×4000 37FF	1 K	I2S2ext
	0×4000 3000 - 0×4000 33FF	1 K	IWDG
	0×4000 2C00 - 0×4000 2FFF	1 K	WWDG
	0×4000 2800 - 0×4000 2BFF	1 K	RTC
	0×4000 1400 - 0×4000 27FF	5 K	Reserved
	0×4000 1000 - 0×4000 13FF	1 K	TIM6
	0×4000 0C00 - 0×4000 0FFF	1 K	Reserved
	0×4000 0800 - 0×4000 0BFF	1 K	Reserved
	0×4000 0400 - 0×4000 07FF	1 K	Reserved
	0×4000 0000 - 0×4000 03FF	1 K	TIM2
	0×2000 A000 - 3FFF FFFF	~512 M	Reserved
	0×2000 0000 - 0×2000 9FFF	40 K	SRAM
	0×1FFF F800 - 0×1FFF FFFF	2 K	Option bytes
	0×1FFF D800 - 0×1FFF F7FF	8 K	System memory
	0×0804 0000 - 0×1FFF D7FF	~384 M	Reserved
	0×0800 0000 - 0×0800 FFFF	64 K	Main Flash memory

Figure 5-4 *AHB1 peripheral allocations.*

GPIO Pin Hardware

Figure 5-5 is a block diagram, which also contains schematics that model a typical STM GPIO port pin.

Starting from the right-hand side of Figure 5-5 you can see the I/O pin, which has robust overvoltage protection circuits attached. The GPIO pin is protected up to an applied 5-V input level in spite of the fact that the main MCU voltage supply is 3.3 V and the core processor voltages are 1.8 V. This makes interfacing quite easy without the need to employ voltage level shifters or resistance voltage dividers. The I/O pin is simultaneously connected to both input and output drivers as shown in the figure. However, only one driver at a time can be activated. Inspecting the input driver section reveals three possible inputs that may be sent to the core MCU. These are as follows:

- Analog: Bypasses the Schmitt trigger
- Alternate function: Schmitt trigger output
- Read: Output from one of the associated read register lines

The GPIO pin configuration settings determine which input is used at a given time.

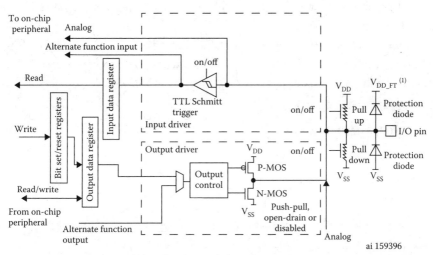

Figure 5-5 *Typical STM GPIO port pin block diagram/schematic.*

Likewise, the output section has multiple ways of creating a signal output for the I/O pin. These are as follows:

- Write: Output from an associated write register line. The output register also has a R/W control line.

- Alternate function: Generated from an internal MCU peripheral.

Note that the output control block has two MOSFET transistors connected to the output control circuitry, which provide for the following operations:

- Push-pull

- Open-drain

- Disabled

Finally, the I/O pin has programmable enabled resistors attached to the pin, which will allow for the following:

- Pull-up

- Pull-down

- Neither

This sample circuitry is replicated for every GPIO pin existing in the MCU.

Each GPIO port has several registers attached to it, which control all of its pins' behaviors by setting or resetting bits associated with the respective pins. Figure 5-6 shows the format for the GPIO port mode register for ports A ... E and H.

This register would be programmatically accessed by using the symbol GPIOx_MODER where x = A ... E and H for the targeted port. The power-on reset values automatically placed in all the GPIO registers are as follows:

- 0x0C00 0000 for port A

- 0x0000 0280 for port B

- 0x0000 0000 for all other ports

The di-bit or two bit values that can be stored in a register have the following meanings:

- 00—Input mode

- 01—Output mode

31	30	29	28	27	26	25	24	23	22	21	20	19	18	17	16
MODER15[1:0]		MODER14[1:0]		MODER13[1:0]		MODER12[1:0]		MODER11[1:0]		MODER10[1:0]		MODER9[1:0]		MODER8[1:0]	
rw	rw	rw	rw	rw	rw	rw	rw	rw	rw	rw	rw	rw	rw	rw	rw
15	14	13	12	11	10	9	8	7	6	5	4	3	2	1	0
MODER7[1:0]		MODER6[1:0]		MODER5[1:0]		MODER4[1:0]		MODER3[1:0]		MODER2[1:0]		MODER1[1:0]		MODER0[1:0]	
rw	rw	rw	rw	rw	rw	rw	rw	rw	rw	rw	rw	rw	rw	rw	rw

Figure 5-6   *GPIO port mode register.*

- 10—Alternate function mode

- 11—Analog mode

Additional GPIO registers further configure other behaviors that are associated with the port pins including the following:

- Enabling push-pull, open-drain, or none (also called floating)

- Enabling pull-up, pull-down, or none

- Operational speed

The next two figures concisely display all the registers involved in all GPIO configurations. These figures are very important and will serve as the basis for configuring GPIO ports for all future projects. I had to break the GPIO register's listings into two separate figures due to the number of involved registers. Figure 5-7 shows the first nine registers in order of their offset addresses to the base MODER address.

The last nine registers in the offset address order are shown in Figure 5-8.

The key to reading these figures is to understand what each register is supposed to control, its offset address, and how individual GPIO pins are accessed in the register.

I will next demonstrate a simple program that will control an LED attached to port A, pin PA10.

LED Test Demonstration

The purpose of the following demonstration is to show how to configure the port A GPIO registers to directly control an LED. The demonstration program will not blink the LED and you have to change the source code, recompile, and reload to affect the LED state.

Offset	Register	31	30	29	28	27	26	25	24	23	22	21	20	19	18	17	16	15	14	13	12	11	10	9	8	7	6	5	4	3	2	1	0
0x00	GPIOA_MODER	MODER15[1:0]		MODER14[1:0]		MODER13[1:0]		MODER12[1:0]		MODER11[1:0]		MODER10[1:0]		MODER9[1:0]		MODER8[1:0]		MODER7[1:0]		MODER6[1:0]		MODER5[1:0]		MODER4[1:0]		MODER3[1:0]		MODER2[1:0]		MODER1[1:0]		MODER0[1:0]	
	Reset value	0	0	0	0	1	1	0	0	0	0	0	0	0	0	0	0	0	0	0	0	0	0	0	0	0	0	0	0	0	0	0	0
0x00	GPIOB_MODER	MODER15[1:0]		MODER14[1:0]		MODER13[1:0]		MODER12[1:0]		MODER11[1:0]		MODER10[1:0]		MODER9[1:0]		MODER8[1:0]		MODER7[1:0]		MODER6[1:0]		MODER5[1:0]		MODER4[1:0]		MODER3[1:0]		MODER2[1:0]		MODER1[1:0]		MODER0[1:0]	
	Reset value	0	0	0	0	0	0	0	0	0	0	0	0	0	0	0	0	0	0	0	0	0	0	0	0	1	0	1	0	0	0	0	0
0x00	GPIOx_MODER (where x = C..E and H)	MODER15[1:0]		MODER14[1:0]		MODER13[1:0]		MODER12[1:0]		MODER11[1:0]		MODER10[1:0]		MODER9[1:0]		MODER8[1:0]		MODER7[1:0]		MODER6[1:0]		MODER5[1:0]		MODER4[1:0]		MODER3[1:0]		MODER2[1:0]		MODER1[1:0]		MODER0[1:0]	
	Reset value	0	0	0	0	0	0	0	0	0	0	0	0	0	0	0	0	0	0	0	0	0	0	0	0	0	0	0	0	0	0	0	0
0x04	GPIOx_OTYPER (where x = A..E and H)	Reserved																OT15	OT14	OT13	OT12	OT11	OT10	OT9	OT8	OT7	OT6	OT5	OT4	OT3	OT2	OT1	OT0
	Reset value																	0	0	0	0	0	0	0	0	0	0	0	0	0	0	0	0
0x08	GPIOx_OSPEEDER (where x = C..E and H)	OSPEEDR15[1:0]		OSPEEDR14[1:0]		OSPEEDR13[1:0]		OSPEEDR12[1:0]		OSPEEDR11[1:0]		OSPEEDR10[1:0]		OSPEEDR9[1:0]		OSPEEDR8[1:0]		OSPEEDR7[1:0]		OSPEEDR6[1:0]		OSPEEDR5[1:0]		OSPEEDR4[1:0]		OSPEEDR3[1:0]		OSPEEDR2[1:0]		OSPEEDR1[1:0]		OSPEEDR0[1:0]	
	Reset value	0	0	0	0	0	0	0	0	0	0	0	0	0	0	0	0	0	0	0	0	0	0	0	0	0	0	0	0	0	0	0	0
0x08	GPIOA_OSPEEDER	OSPEEDR15[1:0]		OSPEEDR14[1:0]		OSPEEDR13[1:0]		OSPEEDR12[1:0]		OSPEEDR11[1:0]		OSPEEDR10[1:0]		OSPEEDR9[1:0]		OSPEEDR8[1:0]		OSPEEDR7[1:0]		OSPEEDR6[1:0]		OSPEEDR5[1:0]		OSPEEDR4[1:0]		OSPEEDR3[1:0]		OSPEEDR2[1:0]		OSPEEDR1[1:0]		OSPEEDR0[1:0]	
	Reset value	0	0	0	0	1	1	0	0	0	0	0	0	0	0	0	0	0	0	0	0	0	0	0	0	0	0	0	0	0	0	0	0
0x08	GPIOB_OSPEEDER	OSPEEDR15[1:0]		OSPEEDR14[1:0]		OSPEEDR13[1:0]		OSPEEDR12[1:0]		OSPEEDR11[1:0]		OSPEEDR10[1:0]		OSPEEDR9[1:0]		OSPEEDR8[1:0]		OSPEEDR7[1:0]		OSPEEDR6[1:0]		OSPEEDR5[1:0]		OSPEEDR4[1:0]		OSPEEDR3[1:0]		OSPEEDR2[1:0]		OSPEEDR1[1:0]		OSPEEDR0[1:0]	
	Reset value	0	0	0	0	0	0	0	0	0	0	0	0	0	0	0	0	0	0	0	0	0	0	0	0	1	1	0	0	0	0	0	0

Figure 5-7 *First nine GPIO registers' details in order of offset addresses.*

However, there is a bit of hardware preparation that must be done before the test program can be run. That involves connecting the LED with a current limiting resistor to the selected port A pin. To ease this requirement, I elected to use an Arduino prototyping board, which is shown in Figure 5-9. This board is quite inexpensive and is readily available from a number of online sources.

Offset	Register	31	30	29	28	27	26	25	24	23	22	21	20	19	18	17	16	15	14	13	12	11	10	9	8	7	6	5	4	3	2	1	0
0x0C	GPIOA_PUPDR	PUPDR15[1:0]		PUPDR14[1:0]		PUPDR13[1:0]		PUPDR12[1:0]		PUPDR11[1:0]		PUPDR10[1:0]		PUPDR9[1:0]		PUPDR8[1:0]		PUPDR7[1:0]		PUPDR6[1:0]		PUPDR5[1:0]		PUPDR4[1:0]		PUPDR3[1:0]		PUPDR2[1:0]		PUPDR1[1:0]		PUPDR0[1:0]	
	Reset value	0	1	1	0	0	1	0	0	0	0	0	0	0	0	0	0	0	0	0	0	0	0	0	0	0	0	0	0	0	0	0	0
0x0C	GPIOB_PUPDR	PUPDR15[1:0]		PUPDR14[1:0]		PUPDR13[1:0]		PUPDR12[1:0]		PUPDR11[1:0]		PUPDR10[1:0]		PUPDR9[1:0]		PUPDR8[1:0]		PUPDR7[1:0]		PUPDR6[1:0]		PUPDR5[1:0]		PUPDR4[1:0]		PUPDR3[1:0]		PUPDR2[1:0]		PUPDR1[1:0]		PUPDR0[1:0]	
	Reset value	0	0	0	0	0	0	0	0	0	0	0	0	0	0	0	0	0	0	0	0	0	0	0	1	0	0	0	0	0	0	0	0
0x0C	GPIOx_PUPDR (where x = C..E and H)	PUPDR15[1:0]		PUPDR14[1:0]		PUPDR13[1:0]		PUPDR12[1:0]		PUPDR11[1:0]		PUPDR10[1:0]		PUPDR9[1:0]		PUPDR8[1:0]		PUPDR7[1:0]		PUPDR6[1:0]		PUPDR5[1:0]		PUPDR4[1:0]		PUPDR3[1:0]		PUPDR2[1:0]		PUPDR1[1:0]		PUPDR0[1:0]	
	Reset value	0	0	0	0	0	0	0	0	0	0	0	0	0	0	0	0	0	0	0	0	0	0	0	0	0	0	0	0	0	0	0	0
0x10	GPIOx_IDR (where x = A..E and H)	Reserved																IDR15	IDR14	IDR13	IDR12	IDR11	IDR10	IDR9	IDR8	IDR7	IDR6	IDR5	IDR4	IDR3	IDR2	IDR1	IDR0
	Reset value																	x	x	x	x	x	x	x	x	x	x	x	x	x	x	x	x
0x14	GPIOx_ODR (where x = A..E and H)	Reserved																ODR15	ODR14	ODR13	ODR12	ODR11	ODR10	ODR9	ODR8	ODR7	ODR6	ODR5	ODR4	ODR3	ODR2	ODR1	ODR0
	Reset value																	0	0	0	0	0	0	0	0	0	0	0	0	0	0	0	0
0x18	GPIOx_BSRR (where x = A..E and H)	BR15	BR14	BR13	BR12	BR11	BR10	BR9	BR8	BR7	BR6	BR5	BR4	BR3	BR2	BR1	BR0	BS15	BS14	BS13	BS12	BS11	BS10	BS9	BS8	BS7	BS6	BS5	BS4	BS3	BS2	BS1	BS0
	Reset value	0	0	0	0	0	0	0	0	0	0	0	0	0	0	0	0	0	0	0	0	0	0	0	0	0	0	0	0	0	0	0	0
0x1C	GPIOx_LCKR (where x = A..E and H)	Reserved															LCKK	LCK15	LCK14	LCK13	LCK12	LCK11	LCK10	LCK9	LCK8	LCK7	LCK6	LCK5	LCK4	LCK3	LCK2	LCK1	LCK0
	Reset value																0	0	0	0	0	0	0	0	0	0	0	0	0	0	0	0	0
0x20	GPIOx_AFRL (where x = A..E and H)	AFRL7[3:0]				AFRL6[3:0]				AFRL5[3:0]				AFRL4[3:0]				AFRL3[3:0]				AFRL2[3:0]				AFRL1[3:0]				AFRL0[3:0]			
	Reset value	0	0	0	0	0	0	0	0	0	0	0	0	0	0	0	0	0	0	0	0	0	0	0	0	0	0	0	0	0	0	0	0
0x24	GPIOx_AFRH (where x = A..E and H)	AFRH15[3:0]				AFRH14[3:0]				AFRH13[3:0]				AFRH12[3:0]				AFRH11[3:0]				AFRH10[3:0]				AFRH9[3:0]				AFRH8[3:0]			
	Reset value	0	0	0	0	0	0	0	0	0	0	0	0	0	0	0	0	0	0	0	0	0	0	0	0	0	0	0	0	0	0	0	0

Figure 5-8   *Last nine GPIO registers' details in order of offset addresses.*

You will also need an LED and a 330-Ω 1/4-watt resistor to limit the current flowing through the LED. It will also be important to know which GPIO pins are exposed in the Arduino protoboard. This is easily answered by referring to the STM Nucleo-64 board user manual. Figure 5-10 comes from that manual and shows the STM32F302R8 pins associated with both the Arduino and Morpho connectors. In our case, only the Arduino pins are needed.

Figure 5-9 *Arduino prototype board.*

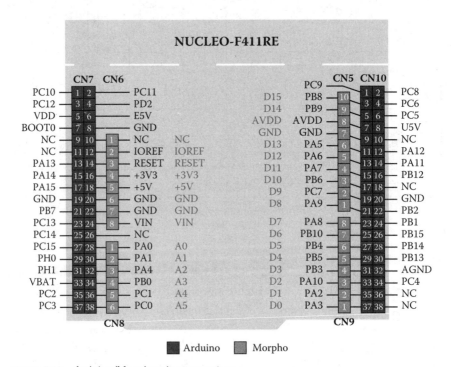

Figure 5-10 *Arduino/Morpho pin connectors.*

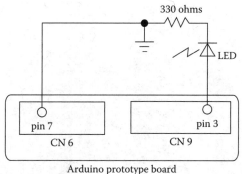

Figure 5-11 *LED test schematic.*

I arbitrarily selected PA10 as the LED output, simply because it was not already configured to a peripheral in the main.c template program. Figure 5-11 is the schematic used for this setup.

The longest LED lead is connected to the anode and the shortest lead is connected to the cathode. I simply soldered the resistor and the LED together because this arrangement allowed me to easily connect the LED/resistor combination to the protoboard, as shown in Figure 5-12. The protoboard is also plugged into the Nucleo-64 board for this photograph.

It is time to work on the software once the hardware is all set.

LED Test Program

As I have previously mentioned, all STM peripherals, including GPIO ports are controlled by configuring a series of related registers, as detailed in Figures 5-7 and 5-8. In this demonstration program I will be configuring pin PA10, which is in the Port A GPIO bank of pins. PA10 will need to be configured as an output. I will be using a slightly modified main method created in the previous chapter's project. You should set up a new project with this main method and simply modify the method by entering the following two lines and making two changes to the `MX_GPIO_Init(void)` method as detailed below.

Enter these two statements immediately after the first `MX_GPIO_Init(}` statement:

```
HAL_GPIO_WritePin(LD2_GPIO_Port, GPIO_PIN_10, GPIO_PIN_
SET);
while (1){}
```

Figure 5-12   *LED/resistor combination plugged into the protoboard.*

Make the following two changes in the static void MX_GPIO_
Init(void) method:

1. HAL_GPIO_WritePin(LD2_GPIO_Port, LED2_Pin, GPIO_PIN_
   RESET);

   to

   HAL_GPIO_WritePin(LD2_GPIO_Port, GPIO_PIN_10, GPIO_
   PIN_RESET);

2. GPIO_InitStruct.Pin = LD2_Pin;

   to

   GPIO_InitStruct.Pin = GPIO_PIN_10;

Structure Component	GPIO Register	Remark
GPIO_InitStruct.Pin	GPIOx_OTYPER	Pin id
GPIO_InitStruct.Mode	GPIOx_MODER	Input or output
GPIO_InitStruct.Pull	GPIOx_PUPDR_	Push-pull, open-drain, floating
GPIO_InitStruct.Speed	GPIOx_OSPEEDER	100, 75, or 50 MHz

Table 5-2 *GPIO_InitStruct to GPIO Register Equivalence*

The HAL library relies on C structure named GPIO_InitStruct to configure any GPIO pin. Using this structure hides the necessity to use actual memory-mapped addresses to configure a GPIO pin. The following code segment from the MX_GPIO_Init(void) method shows how the structure is initialized to configure PA10 to an output pin:

```
GPIO_InitStruct.Pin = GPIO_PIN_10;
GPIO_InitStruct.Mode = GPIO_MODE_OUTPUT_PP;
GPIO_InitStruct.Pull = GPIO_NOPULL;
GPIO_InitStruct.Speed = GPIO_SPEED_FREQ_LOW;
HAL_GPIO_Init(LD2_GPIO_Port, &GPIO_InitStruct);
```

The last list statement is not part of the structure initialization but is required to apply the newly initialized structure within the HAL framework. The structure components are directly related to the memory-mapped GPIO registers, as shown in Table 5-2.

```
/** The STM disclaimer goes here */

// Includes
#include "main.h"
#include "stm32f3xx_hal.h"

/* USER CODE BEGIN Includes */

/* USER CODE END Includes */

// Private variables

/* USER CODE BEGIN PV */
// Private variables

/* USER CODE END PV */

// Private function prototypes
void SystemClock_Config(void);
static void MX_GPIO_Init(void);
```

```
/* USER CODE BEGIN PFP */
// Private function prototypes

/* USER CODE END PFP */

/* USER CODE BEGIN 0 */

/* USER CODE END 0 */

int main(void)
{

  /* USER CODE BEGIN 1 */

  /* USER CODE END 1 */

  // MCU Configuration

  // Reset of all peripherals, Initializes the Flash interface and the
  // Systick.
  HAL_Init();

  /* USER CODE BEGIN Init */

  /* USER CODE END Init */

  /* Configure the system clock */
  SystemClock_Config();

  /* USER CODE BEGIN SysInit */

  /* USER CODE END SysInit */

  /* Initialize all configured peripherals */
  MX_GPIO_Init();

  /* USER CODE BEGIN 2 */
  // The next statement will turn-on GPIOA pin 10
  HAL_GPIO_WritePin(LD2_GPIO_Port, GPIO_PIN_10, GPIO_PIN_SET);

  /* USER CODE END 2 */

  /* Infinite loop */
  /* USER CODE BEGIN WHILE */
  while (1)
  {
  /* USER CODE END WHILE */

  /* USER CODE BEGIN 3 */

  }
  /* USER CODE END 3 */
}
```

```c
// System Clock Configuration
void SystemClock_Config(void)
{
  RCC_OscInitTypeDef RCC_OscInitStruct;
  RCC_ClkInitTypeDef RCC_ClkInitStruct;

  // Initialize the CPU, AHB and APB buss clocks
  RCC_OscInitStruct.OscillatorType = RCC_OSCILLATORTYPE_HSI;
  RCC_OscInitStruct.HSIState = RCC_HSI_ON;
  RCC_OscInitStruct.HSICalibrationValue = 16;
  RCC_OscInitStruct.PLL.PLLState = RCC_PLL_ON;
  RCC_OscInitStruct.PLL.PLLSource = RCC_PLLSOURCE_HSI;
  RCC_OscInitStruct.PLL.PLLMUL = RCC_PLL_MUL16;
  if (HAL_RCC_OscConfig(&RCC_OscInitStruct) != HAL_OK)
  {
    _Error_Handler(__FILE__, __LINE__);
  }

  // Initialize the CPU, AHB and APB buss clocks
  RCC_ClkInitStruct.ClockType =
  RCC_CLOCKTYPE_HCLK|RCC_CLOCKTYPE_SYSCLK
  |RCC_CLOCKTYPE_PCLK1|RCC_CLOCKTYPE_PCLK2;
  RCC_ClkInitStruct.SYSCLKSource = RCC_SYSCLKSOURCE_PLLCLK;
  RCC_ClkInitStruct.AHBCLKDivider = RCC_SYSCLK_DIV1;
  RCC_ClkInitStruct.APB1CLKDivider = RCC_HCLK_DIV2;
  RCC_ClkInitStruct.APB2CLKDivider = RCC_HCLK_DIV1;

  if (HAL_RCC_ClockConfig(&RCC_ClkInitStruct, FLASH_LATENCY_2) !=
  HAL_OK)
  {
    _Error_Handler(__FILE__, __LINE__);
  }

  // Configure the Systick interrupt time
  HAL_SYSTICK_Config(HAL_RCC_GetHCLKFreq()/1000);

  // Configure the Systick
  HAL_SYSTICK_CLKSourceConfig(SYSTICK_CLKSOURCE_HCLK);

  /* SysTick_IRQn interrupt configuration */
  HAL_NVIC_SetPriority(SysTick_IRQn, 0, 0);
}

/** Configure pins as
        * Analog
        * Input
        * Output
        * EVENT_OUT
        * EXTI
    PA2      ------> USART2_TX
    PA3      ------> USART2_RX
*/
```

```
static void MX_GPIO_Init(void)
{
  GPIO_InitTypeDef GPIO_InitStruct;

  /* GPIO Ports Clock Enable */
  __HAL_RCC_GPIOC_CLK_ENABLE();
  __HAL_RCC_GPIOF_CLK_ENABLE();
  __HAL_RCC_GPIOA_CLK_ENABLE();
  __HAL_RCC_GPIOB_CLK_ENABLE();

  // Configure GPIO_PIN_10 pin Output Level */
  HAL_GPIO_WritePin(LD2_GPIO_Port, GPIO_PIN_10, GPIO_PIN_RESET);

  /*Configure GPIO pin : B1_Pin */
  GPIO_InitStruct.Pin = B1_Pin;
  GPIO_InitStruct.Mode = GPIO_MODE_IT_FALLING;
  GPIO_InitStruct.Pull = GPIO_NOPULL;
  HAL_GPIO_Init(B1_GPIO_Port, &GPIO_InitStruct);

  /*Configure GPIO pins : USART_TX_Pin USART_RX_Pin */
  GPIO_InitStruct.Pin = USART_TX_Pin|USART_RX_Pin;
  GPIO_InitStruct.Mode = GPIO_MODE_AF_PP;
  GPIO_InitStruct.Pull = GPIO_NOPULL;
  GPIO_InitStruct.Speed = GPIO_SPEED_FREQ_LOW;
  GPIO_InitStruct.Alternate = GPIO_AF7_USART2;
  HAL_GPIO_Init(GPIOA, &GPIO_InitStruct);

// Configure GPIO pin: GPIO_Pin_10
  GPIO_InitStruct.Pin = GPIO_PIN_10;
  GPIO_InitStruct.Mode = GPIO_MODE_OUTPUT_PP;
  GPIO_InitStruct.Pull = GPIO_NOPULL;
  GPIO_InitStruct.Speed = GPIO_SPEED_FREQ_LOW;
  HAL_GPIO_Init(LD2_GPIO_Port, &GPIO_InitStruct);

}

/* USER CODE BEGIN 4 */

/* USER CODE END 4 */

/**
  * @brief  This function is executed in case of error occurrence.
  * @param  None
  * @retval None
  */
void _Error_Handler(char * file, int line)
{
  /* USER CODE BEGIN Error_Handler_Debug */
  /* User can add his own implementation to report the HAL error
return state */
  while (1)
```

```
    {
    }
    /* USER CODE END Error_Handler_Debug */
}

#ifdef USE_FULL_ASSERT

/**
    * @brief Reports the name of the source file and the source line
number
    * where the assert_param error has occurred.
    * @param file: pointer to the source file name
    * @param line: assert_param error line source number
    * @retval None
    */
void assert_failed(uint8_t* file, uint32_t line)
{
    /* USER CODE BEGIN 6 */
    /* User can add his own implementation to report the file name and
line number,
        ex: printf("Wrong parameters value: file %s on line %d\r\n", file,
line) */
    /* USER CODE END 6 */

}

#endif

/**
    * @}
    */

/**
    * @}
    */

/**** (C) COPYRIGHT STMicroelectronics *****END OF FILE****/
```

You should notice that the main method code listing ends with a `while (1)` statement, which simply forces the MCU to loop forever after the LED is turned on. The only way to turn off the LED is to disconnect the power to the board. Resetting the board will only immediately turn on the LED.

Test Run

You will first need to compile the project in preparation for uploading into the Nucleo-64 board. You then use the ST-LINK utility to transfer the hex file from the project's Debug directory into the Nucleo, as previously described in Chapter 4.

The LED should immediately turn on once the hex file is loaded into the board. Recheck the main.c code if the LED fails to turn-on. Incidentally, recheck that you have inserted the hardware into the proper pin sockets. Sometimes, it is as simple as that to fix a problem.

Enabling Multiple Outputs

It is relatively easy to modify the main.c file to enable multiple outputs. I selected GPIO PA6 as an additional output to control an LED. Incidentally, I referred to the defines contained in the main.h file to determine the pins already committed in the project. I have included the relevant portion of the file in the following listing for your use. Note that PA10 was not added as a define; I would have added that if I were building a permanent program instead of a transient demonstration program. The same is true for pin PA6.

```
#define B1_Pin GPIO_PIN_13
#define B1_GPIO_Port GPIOC
#define USART_TX_Pin GPIO_PIN_2
#define USART_TX_GPIO_Port GPIOA
#define USART_RX_Pin GPIO_PIN_3
#define USART_RX_GPIO_Port GPIOA
#define LD2_Pin GPIO_PIN_5
#define LD2_GPIO_Port GPIOA
#define TMS_Pin GPIO_PIN_13
#define TMS_GPIO_Port GPIOA
#define TCK_Pin GPIO_PIN_14
#define TCK_GPIO_Port GPIOA
#define SWO_Pin GPIO_PIN_3
#define SWO_GPIO_Port GPIOB
```

Make the following changes to enable the multiple LED outputs.
In the `main` method change:

1. `HAL_GPIO_WritePin(LD2_GPIO_Port, GPIO_PIN_10, GPIO_PIN_SET);`

   to

   `HAL_GPIO_WritePin(LD2_GPIO_Port, GPIO_PIN_6 | GPIO_PIN_10, GPIO_PIN_SET);`

In the `static void MX_GPIO_Init(void)` method change:

1. `HAL_GPIO_WritePin(LD2_GPIO_Port, GPIO_PIN_10, GPIO_PIN_RESET);`

   to

   `HAL_GPIO_WritePin(LD2_GPIO_Port, GPIO_PIN_6 | GPIO_PIN_10, GPIO_PIN_RESET);`

2. `GPIO_InitStruct.Pin = GPIO_PIN_10;`

   to

   `GPIO_InitStruct.Pin = GPIO_PIN_6 | GPIO_PIN_10;`

You will need to recompile and upload the new hex file into the Nucleo-64 board to test the new functionality.

Test Run

I left the LED/resistor combination connected between PA10 and ground to recheck that it turned on when the program was loaded. It is always wise to do that type of check because it sometimes happens that unintended things happen when a program modification is done, which is believed to not affect an existing and tested function. I next disconnected the LED/resistor circuit and reconnected it to PA6 and ground and confirmed that it also turned on the LED.

It would be straightforward to enable even more outputs using the same procedure. Just note that you would need to enable another GPIO port bank if you were to use all of the available pins in the GPIOA port.

Push-Button Test Demonstration

This demonstration will show you how to enable a GPIO pin input to detect when a user presses the blue user push button. An LED will light for as long as the push button is pressed. There is no additional hardware required for this demonstration other than what was needed for the previous LED demonstration. I will also be using GPIOA PA10 as an output pin to control the LED.

The `MX_GPIO_Init(void)` method has already configured the blue user push button, so all that is needed is to read the button state in the `main()` method. This is accomplished using the following statement:

`HAL_GPIO_ReadPin(GPIOC, GPIO_PIN_13)`

You should also recall from the Chapter 4 discussion that the blue user push button is connected to GPIO Port C, pin 13 and it is also constantly being pulled high. This means the button state will be high when not pushed and transition to a low value when pushed. The corresponding C snippet must continuously respond to this event as follows:

```
while(1)
{
   if(HAL_GPIO_ReadPin(B1_GPIO_Port, B1_Pin)
   {
      HAL_GPIO_WritePin(LD2_GPIO_Port, GPIO_PIN_10, GPIO_
PIN_RESET);
   }
   else
   {
      HAL_GPIO_WritePin(LD2_GPIO_Port, GPIO_PIN_10, GPIO_
PIN_SET);
   }
}
```

Enter the above code segment in the main() method to continue with the demonstration. Next compile and upload the hex file to the Nucleo-64 board.

Test Run

I observed that the LED turned on for as long as the push button was depressed. Note that you will have to use a small screwdriver to press the button as it is hidden under the Arduino protoboard.

The next demonstration concerns a basic property involved with all MCU signal operations, including the GPIO pins.

Clock Speed Demonstration

Clock speed as related to MCU signal operations is a misnomer. You would ordinarily consider it to be a measure of the maximum number of signal transitions per second. However, it is really a measure of the clock skew or amount of time it takes to transition a signal from 10% to 90% of the maximum value. It also applies in the opposite direction or going from 90% to 10% of the maximum value. The former is known as leading edge delay and the latter is called trailing edge delay.

Clock skew is directly related to true clock speed because the leading and trailing edge delays ultimately determine the maximum possible number of signal transitions per second. I will still refer to clock skew as clock speed because that is how STM choose to label this important signal property.

The important question that one should ask is why should clock speed even be considered. The answer lies in extending both MCU and battery life. The physics behind improving clock speed dictates that it takes energy or power, in this case, to overcome the circuit capacitance effects that mainly determine clock speed. Small amounts of additional power are required to improve the circuit switching times. This implies an increase in current flow from the battery over time and consequently a shortened battery life. In addition, increased power dissipation will induce small amounts of additional heat stress in the MCU and also shorten its life. The aforementioned are the main reasons why clock speed should always be selected to be the lowest level but still support required functionality.

This demonstration will illustrate the actual clock speed timings that are associated with the maximum and minimum clock speed settings. There are four speed settings available with the HAL framework, which are as follows:

- Low
- Medium
- Fast
- Very Fast

I will be using both the Low and Very Fast settings and applying them to signal outputs from pins PC0 and PC1, respectively. No additional MCU hardware is required. However, you will need a fast multi-channel oscilloscope to replicate this demonstration. The oscilloscope should also have a minimum bandwidth rating of at least 100 MHz. I used a Picoscope USB oscilloscope, model 3406B, to capture the measurements. This test device functions with a PC to measure and display up to four independent channels.

Setting the Pin Clock Speeds

The PC0 and PC1 pin clock speeds are set by creating two more GPIO_InitStruct C data structures and placing them in the MX_GPIO_Init(void) method. In addition, there will be a small amount of code required in the main() method.

The two new GPIO_InitStruct C data structures are listed below:

```
// Configure GPIO pin PC0
GPIO_InitStruct.Pin = GPIO_PIN_0;
GPIO_InitStruct.Mode = GPIO_MODE_OUTPUT_PP;
```

```
GPIO_InitStruct.Pull = GPIO_NOPULL;
GPIO_InitStruct.Speed = GPIO_SPEED_FREQ_LOW;
HAL_GPIO_Init(GPIOC, &GPIO_InitStruct);
// Configure GPIO pin PC1
GPIO_InitStruct.Pin = GPIO_PIN_1;
GPIO_InitStruct.Mode = GPIO_MODE_OUTPUT_PP;
GPIO_InitStruct.Pull = GPIO_NOPULL;
GPIO_InitStruct.Speed = GPIO_SPEED_FREQ_VERY_HIGH;
HAL_GPIO_Init(GPIOC, &GPIO_InitStruct);
```

The new code statements to be placed in the main() method are as follows:

```
while (1) {
    GPIOC->ODR = 0x3;
    GPIOC->ODR = 0x0;
}
```

The new C structs set up the pins as identical GPIO outputs with the exception that one is low speed and the other is very fast speed. The code segment in main() simply toggles the pin outputs between high and low at the fastest possible rate because there is no other intervening code between the output statements.

Test Results

Figure 5-13 is a screenshot showing an instantaneous capture of the GPIO pin signal waveforms.

It is easy to see that the signal waveforms are not crisp digital signals with sharp transitions but instead are slightly distorted with definite rising and falling edge

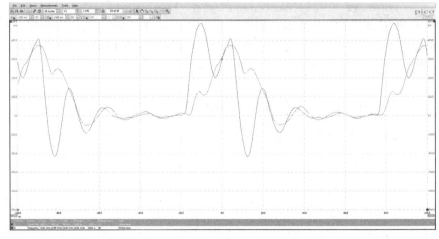

Figure 5-13 *PC0 and PC1 pins' signal waveforms.*

slopes. The distortions are likely due to high oscilloscope probe capacitive loading and the slight inductance due to the jumper leads I connected between the oscilloscope probes and the Arduino protoboard sockets. I also used 10× high impedance probe settings to help reduce the loading on the GPIO pins. I know this helped because I repeated the measurements using the direct X1 probe settings, which further distorted the settings. The vertical voltage amplitude scale must be multiplied by 10 to account for the 10× probe settings.

If you carefully examine the figure you should see two automatic measurements that were computed by the oscilloscope on the displayed waveforms. I have repeated these measurements below for easy viewing:

Pulse period: 94.1 ns

Frequency: 10.62 MHz

I also closely examined waveforms in Figure 5-13 to estimate parameters that I used to calculate the equivalent bandwidth necessary reflecting the measured waveforms leading edges. I also used a rule of thumb for estimating the equivalent bandwidth using the leading edge rise time. These calculations follow:

Low speed:

Leading edge rise time for 10% to 90% of peak = 18 ns

BW (GHz) = 0.35/rise time (nS) = 0.35/18 = 0.019.4

BW (MHz) = 19.4

Very Fast speed:

Leading edge rise time for 10% to 90% of peak = 5 ns

BW (GHz) = 0.35/rise time (nS) = 0.35/5 = 0.070

BW (MHz) = 70.0

The 70-MHz bandwidth calculated for the Very Fast speed setting reflects several factors including the maximum possible AHB1 bus speed, which is 100 MHz and the STM32F302R8 high-speed clock rate, which is 72 MHz for the Nucleo-64 development board it controls. All-in-all, a 70-MHz bandwidth is fairly decent while the 19.4 MHz for the Low speed setting could be problematic for demanding applications such as very high-speed data communications. My experimental results actually agree quite closely with STM's estimates of 75 MHz for the Very Fast setting and 25 MHz for the Low setting.

Summary

This chapter's focus was on how to work with GPIO pins and the HAL. I began with a review of how STM peripherals are memory-mapped because that is the fundamental basis on how the MCU configures and controls them.

I next proceeded to discuss how a typical GPIO hardware pin is set up and the various functions by which it may be configured including as an input, an output, an analog role, or an alternate function. The various control registers associated with a GPIO pin were also discussed.

A simple LED test demonstration was next shown where I used an LED/resistor combination, which was inserted into Arduino protoboard sockets. The protoboard in turn was inserted into the Nucleo-64 development board. I went through all the program changes required to be entered into a main.c file in order to light the LED. A complete and annotated main.c listing was provided for your reference and use.

The above program was next modified to accommodate multiple outputs, which controlled an additional LED.

The next demonstration illustrated how the blue user bush button located on the Nucleo-64 development board could be programmed with a GPIO input pin to light an LED when it was pressed. The purpose of this program was to show you how to set up a GPIO pin as an input.

The last demonstration concerned how the GPIO clock speed parameter affected GPIO signal output waveforms. I explained that the clock speed was a misnomer and the real intent was to control GPIO signal skew. Skew is the time it takes to transition from one state to another state.

6

Interrupts

This chapter covers interrupts. As the name implies, they deal with stopping the normal program execution upon recognition of a predesignated event and then attending to that event.

Interrupts

An interrupt from a high-level perspective is any process that momentarily stops an ongoing program execution and requires the MCU to execute specialized code related to the event that initiated the interrupt. This specialized code is known as an interrupt handler. The event that causes the interrupt is termed an *asynchronous event*, meaning that it has no causal relationship with the normally executing program code. You will shortly see that timer peripherals are typically the source of these asynchronous events. Data communication channels as well as direct memory access (DMA) processes also commonly use interrupts.

An interrupt can originate with hardware, which is the case when a timer is involved. It can also be purely software driven, which will often happen when an abnormal program termination is detected, such as an attempt to divide by zero or attempting to access a nonexistent memory location. Software interrupts in most computer languages are normally referred to as exceptions. However, in ARM terminology, all interrupts whether they are hardware or software initiated are called exceptions.

The ARM Cortex-M integrated peripheral (IP) contains a hardware device called the nested vector interrupt controller (NVIC) whose purpose is to manage exceptions. Figure 6-1 is a block diagram illustrating how the NVIC is connected with other key Cortex-M components, including the Cortex-M core and other peripherals.

You can see two types of wired connections in the left side of Figure 6-1. A single line labeled NMI stands for non-maskable interrupt and is used to allow a peripheral to signal the NVIC to unconditionally start the interrupt process. The other interrupt type is IRQ, which stands for interrupt request. This interrupt type can be masked, meaning that the NVIC does not have to immediately respond to an IRQ if a corresponding masking IRQ register bit has been set. There is usually only one NMI, but there are many IRQ lines that are dependent on the specific model NVIC capacity.

The NVIC is also linked between the Cortex-M core with both dedicated I/O lines and the standard internal bus interconnect structure. This means the NVIC is completely CMCIS compliant and uses the same type of control and configuration registers as has been detailed in the GPIO discussion. The NVIC is fully integrated into the HAL software framework and uses its own C structures to set up a variety of registers. Driver software for the NVIC will be discussed in a demonstration program. At this point, I will discuss some important details regarding the NVIC.

NVIC Specifications

The STM NVIC general specification states that up to 256 interrupt channels can be managed by a single NVIC. Out of the 256 channels, 240 may be external, while 16 are always internal. In reality, the NVIC portion of the STMF302R8 MCU, which controls the project development board, only supports 52 channels. In

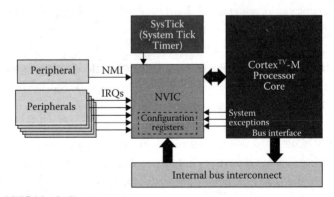

Figure 6-1 *NVIC block diagram.*

addition, there are 16 hardware interrupt lines that have direct connections to the NVIC. Any of the MCU's 51 GPIO lines may be configured as a maskable interrupt, which means they can become an interrupt channel. You should think of an interrupt channel as more of a software construct than a hardware implementation. In fact, the interrupt C structure uses the word "channel" to reinforce this concept.

Interrupts also may be assigned a priority level. This means that a higher priority interrupt can interrupt a lower priority interrupt. This action gives meaning to the word "nested" in the acronym NVIC. An 8-bit register holds interrupt priority levels, which means there can be a maximum of 256 priority levels. I cannot conceive of a practical embedded system that would ever need anywhere that many interrupt priority levels. Such a system would likely be impossible to design, let alone implement.

It is time to now return to a detailed discussion of the interrupt process as I have already introduced the NVIC.

Interrupt Process

There a few definitions to be set forth before proceeding with this section:

- Arm: Allows a hardware device to trigger an interrupt
- Disarm: Disallows a device to trigger an interrupt
- Enable: Allows the interrupt process
- Disable: Disallows the interrupt process

Enabling and disabling interrupts are easily accomplished in the C language by using the `EnableInterrupts()` and `DisableInterrupts()` methods, respectively. A potential interrupt source can trigger an interrupt provided that interrupts are enabled and a specific trigger bit associated with that potential source has been set in the appropriate NVIC control register. This is considered arming a specific device. Conversely, resetting that same register trigger bit will be considered disarming the device. There is one important deviation from this procedure when considering the SysTick interrupt. It has no enable bit because of its importance in having its events always recognized by the MCU.

Interrupt priorities are established by setting appropriate bits in the NVIC BASEPRI register. For example, if a specific interrupt is set at a priority level equal to 3, then any interrupt with priority level of 0, 1, or 2 will interrupt the level 3 interrupt. However, if the BASEPRI register is set to 0, then the interrupt priority feature is disabled and all interrupts are processed as they are received by the NVIC.

There are five preconditions that must be true before an interrupt can be generated:

1. The potential interrupt device must be armed, that is, trigger bit set.

2. NVIC must be enabled.

3. Global interrupts must be enabled.

4. Interrupt priority must be higher than the current level.

5. An actual hardware triggering event must happen.

These five conditions can happen in any order. However, they must all be true simultaneously for an interrupt to be recognized and processed.

The following five events will happen if an interrupt is recognized:

1. The instruction currently executing will complete.

2. The currently operating thread will be suspended.

3. The contents of 12 registers will be stored on the stack. An additional 18 words will be stored if the floating point unit (FPU) happened to be active when the interrupt was recognized.

4. The LR register is set to a specific value. (LR is a special stack register.)

5. The PC register is loaded with the interrupt service routine (ISR) address.

Figure 6-2 is a graphical depiction of the five events listed above.

The key point to make is that the NVIC upon recognizing a valid interrupt request will interface automatically with the MCU using firmware stored within the NVIC. No special code is needed, which makes the interrupt processing as fast as possible. It only takes 12 clock cycles, as shown in Figure 6-2, to go from recognizing an interrupt to executing the first instruction in the ISR. This means there is a latency time of approximately 1.5 μs for an MCU clocked at 80 MHz.

This series of events is called a context switch and is automatically invoked by the NVIC hardware. The result of a context switch is that the currently operating thread becomes a background thread and the ISR becomes a foreground thread. The context switch also works quite nicely with prioritized interrupts where the lower priority ISR becomes a background thread and the higher priority ISR becomes a foreground thread.

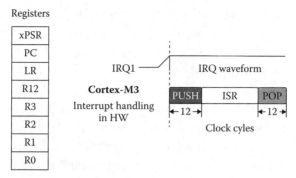

Figure 6-2 *Interrupt processing sequence.*

There is also the concept of a pending interrupt request, which you should know about. A requesting interrupt is held in a pending state if it is detected if the global interrupts are not yet enabled or the requesting interrupt has a lower priority than a currently active interrupt. Once the five preconditions are met, the requesting interrupt's state will be changed from pending to active.

It is also very important to clear all device trigger bits or else there will be continuous interrupt requests reoccurring. It is called an acknowledgment to clear a trigger bit and is done by a software instruction. The only device which has an automatic acknowledgment is the SysTick interrupt because of its critical role in MCU operation and its recurring nature of operation. The acknowledgment instruction is almost always included in the ISR.

A few words on the nature of an ISR would be helpful in order to understand what role it plays in the interrupt process. The ISR contains all the code required to process and complete whatever action prompted the initial interrupt. An ISR is usually a very compact software module designed for a single task supporting the interrupting device. However, there is one implementation where just one large ISR services many interrupt sources. In this case, a polling technique is employed to determine the exact interrupt source and use just that portion of the large ISR to service the interrupting device. This is not the case for STM MCUs. The STM implementation uses separate and small ISRs to service each requesting interrupt device. There is a list of addresses for these small ISRs stored in a vector table. This feature gives rise to the word "vector" in the NVIC name.

I do want to mention one area where the STM vectored interrupt process actually uses polling. This would be the case where multiple GPIO pins on the same port could separately issue their own interrupt. Since the GPIO port only has one vector address, it is left to the software to poll all the port's trigger bits to determine which bit is armed (requested the interrupt).

Every ISR is written to service the interrupting device as efficiently as possible and will issue the acknowledgment as well as invoke the context switch. The latter is easily accomplished by a C language command that has an assembly language format similar to BX LR, which will very quickly pop all the stored register values back to their respective registers and thus restore the MCU to its pre-interrupt state. It is software engineering design objective that ISRs should run as quickly as possible and return to the previous foreground thread. This means no wait loops should ever be placed in an ISR.

Any data which is generated or modified in an ISR must be stored in globally shared memory locations, because it will otherwise not be available to any other software module due to the nature of how the C language implements the ISR framework.

External Interrupts

External interrupts such as those which are sourced through GPIO pins are connected to the NVIC using an external interrupt controller (EXTI). Using an EXTI allows for multiplexing of many GPIO pins into the NVIC external interrupt inputs. Figure 6-3 is a block diagram of the EXTI.

The EXTI controller in the STM32F302R8 MCU carries out the following activities:

- Handles up to 23 separate software events and/or hardware triggers connected to 16 external interrupt/event lines

- Provides a status bit for each line

- Provides independent triggers and masks for each line

- Can detect external pulse with a shorted clock period than the APB2 bus clock, which drives the EXTI

Any of the STM32F302R8 51 GPIO pins may be used as an interrupt source. Of course, only up to 23 could be active at any given time. That many external interrupts should easily be able to handle any conceivable embedded design. There are also five other EXTI lines that are connected as follows:

1. EXTI line 16 connected to the PVD output

2. EXTI line 17 connected to the RTC Alarm event

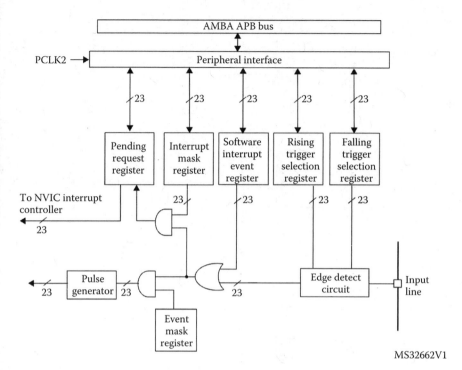

Figure 6-3 *EXTI block diagram.*

3. EXTI line 18 connected to the USB OTG FS Wakeup event

4. EXTI line 21 connected to the RTC Tamper and TimeStamp events

5. EXTI line 22 connected to the RTC Wakeup

Figure 6-4 shows relationship between the GPIO pins, external interrupt controller inputs, and the NVIC IRQs.

You should take note that GPIO pin PC13 is grouped together with many other pins into a common NVIC IRQ named EXTI15_10_IRQ. This is quite important in the interrupt program demonstration.

At this point, you should have gained sufficient knowledge to appreciate a simple interrupt demonstration discussed in the next section.

Interrupt Demonstration

This interrupt demonstration will be created such that a user can push the blue button on the Nucleo-64 board and cause the user LED to blink at a different

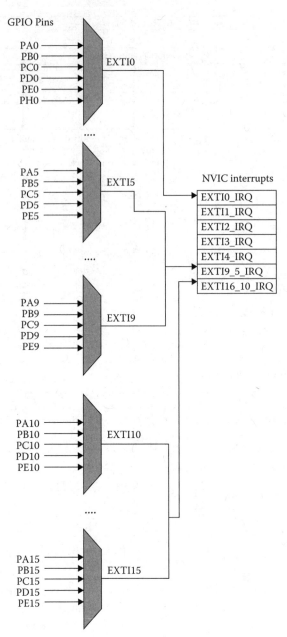

Figure 6-4 *GPIO pins, external interrupt controller, and NVIC IRQs.*

interval toggle, that is, push it once to blink twice per second turn and push it again to have it blink once every two seconds. The program takes full advantage of the HAL framework to invoke the appropriate GPIO interrupt for GPIO pin PC13 connected to the blue user push button. The ISR contains code that toggles LD2, a green LED, which is connected to GPIO pin PA5.

The following is a complete code listing for the main.c file contained in the project. A discussion of key parts of the code follows the listing.

```c
// STM disclaimer goes here
// Includes
#include "main.h"
#include "stm32f3xx_hal.h"

/* USER CODE BEGIN Includes */

/* USER CODE END Includes */

// Private variables

/* USER CODE BEGIN PV */
// Private variables
uint16_t tDelay = 250;  // also used as an external variable
/* USER CODE END PV */

// Private function prototypes
void SystemClock_Config(void);
static void MX_GPIO_Init(void);

/* USER CODE BEGIN PFP */
// Private function prototypes

/* USER CODE END PFP */

/* USER CODE BEGIN 0 */

/* USER CODE END 0 */

int main(void)
{

  /* USER CODE BEGIN 1 */

  /* USER CODE END 1 */

  // MCU Configuration

  // Reset of all peripherals, Initialize the flash and the Systick.
  HAL_Init();
```

```
/* USER CODE BEGIN Init */

/* USER CODE END Init */

/* Configure the system clock */
SystemClock_Config();

/* USER CODE BEGIN SysInit */

/* USER CODE END SysInit */

/* Initialize all configured peripherals */
MX_GPIO_Init();

/* USER CODE BEGIN 2 */

/* USER CODE END 2 */

/* Infinite loop */
/* USER CODE BEGIN WHILE */
while (1)
{
/* USER CODE END WHILE */

/* USER CODE BEGIN 3 */
HAL_GPIO_TogglePin(LD2_GPIO_Port, LD2_Pin);
HAL_Delay(tDelay);

}
/* USER CODE END 3 */

}

// System Clock Configuration
void SystemClock_Config(void)
{

  RCC_OscInitTypeDef RCC_OscInitStruct;
  RCC_ClkInitTypeDef RCC_ClkInitStruct;

  // Initialize the CPU, AHB and APB buss clocks
  RCC_OscInitStruct.OscillatorType = RCC_OSCILLATORTYPE_HSI;
  RCC_OscInitStruct.HSIState = RCC_HSI_ON;
  RCC_OscInitStruct.HSICalibrationValue = 16;
  RCC_OscInitStruct.PLL.PLLState = RCC_PLL_ON;
  RCC_OscInitStruct.PLL.PLLSource = RCC_PLLSOURCE_HSI;
  RCC_OscInitStruct.PLL.PLLMUL = RCC_PLL_MUL16;
  if (HAL_RCC_OscConfig(&RCC_OscInitStruct) != HAL_OK)
  {
    _Error_Handler(__FILE__, __LINE__);
  }
```

```
  // Initializes the CPU, AHB and APB buss clocks
  RCC_ClkInitStruct.ClockType =
  RCC_CLOCKTYPE_HCLK|RCC_CLOCKTYPE_SYSCLK
  |RCC_CLOCKTYPE_PCLK1|RCC_CLOCKTYPE_PCLK2;
  RCC_ClkInitStruct.SYSCLKSource = RCC_SYSCLKSOURCE_PLLCLK;
  RCC_ClkInitStruct.AHBCLKDivider = RCC_SYSCLK_DIV1;
  RCC_ClkInitStruct.APB1CLKDivider = RCC_HCLK_DIV2;
  RCC_ClkInitStruct.APB2CLKDivider = RCC_HCLK_DIV1;

  if (HAL_RCC_ClockConfig(&RCC_ClkInitStruct, FLASH_LATENCY_2) !=
HAL_OK)
  {
    _Error_Handler(__FILE__, __LINE__);
  }

  // Configure the Systick interrupt time
  HAL_SYSTICK_Config(HAL_RCC_GetHCLKFreq()/1000);

  // Configure the Systick
  HAL_SYSTICK_CLKSourceConfig(SYSTICK_CLKSOURCE_HCLK);

  /* SysTick_IRQn interrupt configuration */
  HAL_NVIC_SetPriority(SysTick_IRQn, 0, 0);
}
/** Configure pins as
        * Analog
        * Input
        * Output
        * EVENT_OUT
        * EXTI
     PA2     ------> USART2_TX
     PA3     ------> USART2_RX
*/
static void MX_GPIO_Init(void)
{

  GPIO_InitTypeDef GPIO_InitStruct;

  /* GPIO Ports Clock Enable */
  __HAL_RCC_GPIOC_CLK_ENABLE();
  __HAL_RCC_GPIOF_CLK_ENABLE();
  __HAL_RCC_GPIOA_CLK_ENABLE();
  __HAL_RCC_GPIOB_CLK_ENABLE();

  /*Configure GPIO pin Output Level */
  HAL_GPIO_WritePin(LD2_GPIO_Port, LD2_Pin, GPIO_PIN_RESET);

  /*Configure GPIO pin : PC13 */
  GPIO_InitStruct.Pin = GPIO_PIN_13;
  GPIO_InitStruct.Mode = GPIO_MODE_IT_RISING;
```

```
GPIO_InitStruct.Pull = GPIO_NOPULL;
HAL_GPIO_Init(GPIOC, &GPIO_InitStruct);

/*Configure GPIO pins : USART_TX_Pin USART_RX_Pin */
GPIO_InitStruct.Pin = USART_TX_Pin|USART_RX_Pin;
GPIO_InitStruct.Mode = GPIO_MODE_AF_PP;
GPIO_InitStruct.Pull = GPIO_NOPULL;
GPIO_InitStruct.Speed = GPIO_SPEED_FREQ_LOW;
GPIO_InitStruct.Alternate = GPIO_AF7_USART2;
HAL_GPIO_Init(GPIOA, &GPIO_InitStruct);

/*Configure GPIO pin : LD2_Pin */
GPIO_InitStruct.Pin = LD2_Pin;
GPIO_InitStruct.Mode = GPIO_MODE_OUTPUT_PP;
GPIO_InitStruct.Pull = GPIO_NOPULL;
GPIO_InitStruct.Speed = GPIO_SPEED_FREQ_LOW;
HAL_GPIO_Init(LD2_GPIO_Port, &GPIO_InitStruct);

/* EXTI interrupt init*/
HAL_NVIC_SetPriority(EXTI15_10_IRQn, 0, 0);
HAL_NVIC_EnableIRQ(EXTI15_10_IRQn);

}

/* USER CODE BEGIN 4 */

/* USER CODE END 4 */

/**
  * @brief  This function is executed in case of error occurrence.
  * @param  None
  * @retval None
  */
void _Error_Handler(char * file, int line)
{
  /* USER CODE BEGIN Error_Handler_Debug */
  /* User can add his own implementation to report the HAL error re-
turn state */
  while(1)
  {
  }
  /* USER CODE END Error_Handler_Debug */
}

#ifdef USE_FULL_ASSERT

/**
  * @brief Reports the name of the source file and the source line
number
  * where the assert_param error has occurred.
  * @param file: pointer to the source file name
```

```
   * @param line: assert_param error line source number
   * @retval None
   */
void assert_failed(uint8_t* file, uint32_t line)
{
  /* USER CODE BEGIN 6 */
  /* User can add his own implementation to report the file name and
line number,
     ex: printf("Wrong parameters value: file %s on line %d\r\n", file,
line) */
  /* USER CODE END 6 */

}

#endif

/**
  * @}
  */

/**
  * @}
  */

/********************** (C) COPYRIGHT STMicroelectronics *****END OF
FILE****/
```

Examining the `MX_GPIO_Init` method reveals that it is not much different than the code for the same method used in the previous chapter. The two changes in the method are as follows:

GPIO pin PC13 has been configured to generate an interrupt when pressed by the following statement:

```
GPIO_InitStruct.Mode = GPIO_MODE_IT_RISING;
```

Next, the external interrupt was assigned a priority by the statement:

```
HAL_NVIC_SetPriority(EXTI15_10_IRQn, 0, 0);
```

The NVIC was then enabled to recognize an interrupt from PC13 by the following statement:

```
HAL_NVIC_EnableIRQ(EXTI15_10_IRQn);
```

That's all the changes necessary to be applied to the `MX_GPIO_Init` method.

The only thing left is to add the ISR handler code to the project. This ISR must be placed in the stm32f3xx_it.c file. The ISR signature is void

EXTI15_10_IRQHandler(void). The complete stm32f3xx_it.c file listing is shown below:

```c
// STM disclaimer goes here
// Includes
#include "stm32f3xx_hal.h"
#include "stm32f3xx.h"
#include "stm32f3xx_it.h"

/* USER CODE BEGIN 0 */
/* USER CODE END 0 */

// External variables
extern uint16_t tDelay;

/**
* Cortex-M4 Processor Interruption and Exception Handlers
*/

/**
* @brief This function handles System tick timer.
*/
void SysTick_Handler(void)
{
  /* USER CODE BEGIN SysTick_IRQn 0 */

  /* USER CODE END SysTick_IRQn 0 */
  HAL_IncTick();
  HAL_SYSTICK_IRQHandler();
  /* USER CODE BEGIN SysTick_IRQn 1 */

  /* USER CODE END SysTick_IRQn 1 */
}

/**
* STM32F3xx Peripheral Interrupt Handlers
* Add here the Interrupt Handlers for the used peripherals.
*
* For the available peripheral interrupt handler names,
* please refer to the startup file (startup_stm32f3xx.s).
*/

/**
* @brief This function handles RCC global interrupt.
*/
void RCC_IRQHandler(void)
{
  /* USER CODE BEGIN RCC_IRQn 0 */

  /* USER CODE END RCC_IRQn 0 */
  /* USER CODE BEGIN RCC_IRQn 1 */
```

```
  /* USER CODE END RCC_IRQn 1 */
}

/**
* @brief This function handles EXTI line[15:10] interrupts.
*/
void EXTI15_10_IRQHandler(void)
{
/* USER CODE BEGIN EXTI15_10_IRQn 0 */
    for(int i = 0; i < 65535; i++); // short delay for debounce
        if(HAL_GPIO_ReadPin(GPIOC, GPIO_PIN_13)) // no defines used
        {
            if(tDelay == 250)
                tDelay = 1000;
            else
                tDelay = 250;
        }
/* USER CODE END EXTI15_10_IRQn 0 */

  HAL_GPIO_EXTI_IRQHandler(GPIO_PIN_13);
  /* USER CODE BEGIN EXTI15_10_IRQn 1 */

  /* USER CODE END EXTI15_10_IRQn 1 */
}

/* USER CODE BEGIN 1 */

/* USER CODE END 1 */
/**** (C) COPYRIGHT STMicroelectronics *****END OF FILE****/
```

You should immediately see that the ISR has only seven statements, which agrees very nicely with my earlier discussion that ISRs must be as concise as possible and dedicated to a single task. The first statement is a no operation (nop) "for" loop that introduces a very short delay to debounce the push button. The next statement is a compound "if" that reads the voltage level connected to the push button. It will be high because the program is already in the interrupt handler. If the external C variable tDelay is set to 250, it will be reset to 1000 and vice versa. The last statement in the ISR activates the NVIC IRQ handler for the external controller connected to pin PC13.

Test Run

I built the project and downloaded that into the Nucleo-64 board using the procedure detailed multiple time in previous chapters. I initially observed that the LED blinked twice a second after the initial download. I then pressed the blue button and observed that the LED started blinking at a rate of once every two seconds. Pressing it again returned to the twice a second blink rate. This was a convincing proof that the interrupt demonstration was working as desired. Admittedly, this is a very simple demonstration, yet it contains all the elements necessary to support far more complex interrupt driven programs. You find that interrupts are an embedded developer's best friend. Most of my embedded development projects use interrupts. Not using them deprives you of an important tool in creating efficient embedded programs. It is not an overstatement to say that modern embedded projects would not be practical without employing at least one interrupt.

Summary

I started the chapter by discussing how the STM MCU interrupt process functions. This discussion included the description and functioning of the NVIC and a walk through a complete interrupt processing sequence.

An interrupt demonstration next followed in which a user pressed the blue push button located on the project Nucleo-64 board, which caused an interrupt to be generated. The corresponding interrupt service routine (ISR) code consequently toggled the state of an onboard LED.

7

Timers

Timer peripherals are very important components within a MCU. Many embedded applications are time or temporal dependent and timers are the primary means by which the MCU controls the application. While it is certainly possible to use a MCU to directly time processes, it would be great waste of processing power and highly inefficient approach. Using hardware timers along with interrupts is really the only practical way to implement embedded time-dependent applications. Fortunately, there is a powerful timer variety provided by STM in its line of MCUs.

STM Timer Peripherals

A timer is simply a free-running counter that counts pulses from a clock source. Since the pulse train from a clock source has a known period or interval between pulses, the elapsed time is directly related to the number of pulses counted. The timer clock source used with a typical MCU is usually derived from an internal master clock. However, the master clock usually runs at a very fast speed, so timer pulses must be divided by hardware to more reasonable values that can be used by timers. The hardware used to "slow" the master clock rate varies and can be binary dividers (prescaler) or phase-locked loops (PLL). Timer counters can count up or accumulate counts until they overflow based on the number of bits used in the counter. A 16-bit counter will overflow when the maximum count of 65535 is reached. An overflow interrupt is often the event which happens with this situation.

Conversely, a timer counter can count down from some preset value and trigger an interrupt when it reaches a 0 value. This counter type is known as a decrementing counter and has many uses.

Timers have many uses, including but not limited to the following:

- Generating a precise time base. All STM timers can do this.

- Measuring the frequency of an incoming digital pulse train.

- Measuring elapsed time on an output signal. This is called output compare for STM timers.

- Generating precise pulse-width modulation (PWM) signals used for servo and motor control.

- Generating single pulse with programmable length and delay characteristics.

- Generating periodic direct memory access (DMA) signals in response to update, trigger, input capture, and output compare events.

There are following five broad categories of STM timers:

- Basic: Simple 16-bit timers. They do not have inputs or output pins. They are principally used as "masters" for other timers. They also are used as a digital-to-analog converter (DAC) clock source. They can generate a time base like all STM timers can do.

- General purpose (GP): These are 16-bit or 32-bit timers with both input and output pins. They can do all of the functions described in the above list. GP timers can have up to four programmable channels, which may be configured as follows:

  - One or two channels

  - One or two channels with a complementary output. The complementary output has a "dead time" generator, which provides for an independent time base.

- Advanced: All the features of the GP timer with additional functions related to motor control and digital power conversion. There are three complementary outputs available with this timer category with an emergency shutdown input.

- High resolution: Multiple high-resolution outputs made possible by six sub-timers, one master and five slaves. Multiple dead-time insertions are possible with this timer. There are also five fault inputs and 10 external event inputs. This timer has the acronym HRTIM1 in STM terminology.

Timer Type		Name
Advanced		TIM1
GP	16-bit	TIM15 TIM16 TIM17
GP	32-bit	TIM2
Basic		TIM6

Table 6-1 *STM32F302R8 MCU Timers*

- Low power: Timers used in exceptional low-power applications. These timers are designed to stay running in just about all STM MCU modes, except the Standby mode. They can even continue to operate without an internal clock source.

The STM32F302R8 MCU used in the demonstration board has basic, GP, and advanced timers. I will focus on how these timers function and are programmed. The concepts discussed will also apply to the other timer types if you encounter them in other projects. As always, the manufacturer datasheets are your best resource to determine how a particular timer works and how it should be programmed.

The STM32F302R8 MCU has six timers, as shown in Table 6-1. This table has a handy reference to keep in mind when selecting and programming a timer.

I will now discuss how to configure a STM timer using the HAL framework.

STM Timer Configuration

STM timers are configured in the same manner as GPIO pins and interrupts as I have previously discussed. STM timers are configured with a C structure named TIM_HandleTypeDef, which contains the following members:

```
uint32_t Prescaler;
uint32_t CounterMode;
uint32_t Period;
uint32_t ClockDivision
uint32_t RepetitionCounter;
```

The struct members represent the following timer parameters and settings:

- Prescaler: The division factor used to scale the master clock rate. There are only 16-bit prescalers used in all the timers. A 16-bit register can hold prescaler values ranging from 1 to 65535. For example, a prescaler value of 40,000 applied to an 80-MHz master clock would mean a 2-kHz clock rate would be input into the timer.

- CounterMode: Sets the count direction. The available CounterModes are as follows:

  - TIM_COUNTERMODE_UP
  - TIM_COUNTERMODE_DOWN
  - TIM_COUNTERMODE_CENTERALIGNED1
  - TIM_COUNTERMODE_CENTERALIGNED2
  - TIM_COUNTERMODE_CENTERALIGNED3

- Period: This is a number that represents the maximum time to be elapsed before the timer counter is reloaded with this number. The maximum number is 0xffff for 16-bit counters and 0xffff ffff for 32-bit counters. A value of 0x0 will ensure a timer does not start.

- ClockDivision: This is a bit-specific field used for setting the ratio between the internal timer clock frequency and a sampling clock used for digital filters. It is also used to set dead-time parameters. The available ClockDivision modes are as follows:

  - TIM_CLOCKDIVISION_DIV1
  - TIM_CLOCKDIVISION_DIV2
  - TIM_CLOCKDIVISION_DIV4

- RepetitionCounter: Sets a limit for the number of times a timer can overflow or underflow. The timer update register will be set when the limit is reached. An event can also be raised in conjunction with the register update.

Update Event Calculation

The following equation should be used to compute the time between update events for a given high-speed clock rate, a prescalar value, and a period value:

$$Update\ event = \frac{High\ speed\ clock}{(Prescaler + 1)(Period + 1)}$$

A sample calculation is shown below:

Given:

   High-speed clock = 80 MHz

   Prescaler = 39999

   Period = 1999

$$Update\ event = \frac{80000000}{(39999+1)(1999+1)} = \frac{80000000}{80000000} = 1.0\ s$$

The discussion on the demonstration project presented below will show you how to configure a GP timer in a polled mode, which in turn will blink an LED.

Polled or Non-interrupt Blink LED Timer Demonstration

This is the first of two demonstrations showing you how to configure the TIM2 GP timer to control LD2, the onboard green LED. This demonstration will not use an interrupt, so I can focus on the discussion of how to configure a timer. The LED will blink at a 0.5-s rate based on the selection of prescaler and period values as well as the Nucleo-64 board's preset 64-MHz high-speed clock rate.

You should start this project using the CubeMX application by providing an appropriate project name. I used the name TIM2_Example1 for this project. In the Pinout view, you should select the internal high-speed clock as timer TIM2 clock source, which is shown in Figure 7-1.

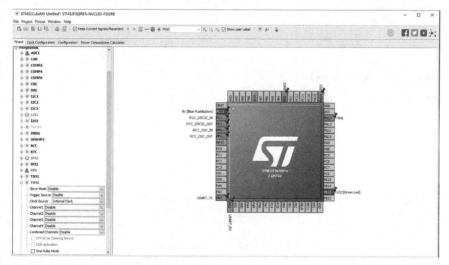

Figure 7-1 *TIM2 clock source selection.*

Selecting the clock source also triggers the CubeMX application to include all the TIM2 initialization statements to be included in the new project. This step is vital to generate a working project.

Don't forget that all the other previous project creation steps are still appropriate, such as including only the necessary library files and ensuring that a hex file is generated.

I will discuss the key parts of the project code after the below listing.

```
// STM disclaimer goes here
// Includes
#include "main.h"
#include "stm32f3xx_hal.h"

/* USER CODE BEGIN Includes */

/* USER CODE END Includes */

// Private variables
TIM_HandleTypeDef htim2;

/* USER CODE BEGIN PV */
// Private variables

/* USER CODE END PV */

// Private function prototypes
void SystemClock_Config(void);
static void MX_GPIO_Init(void);
static void MX_TIM2_Init(void);

/* USER CODE BEGIN PFP */
// Private function prototypes

/* USER CODE END PFP */

/* USER CODE BEGIN 0 */

/* USER CODE END 0 */

int main(void)
{

  /* USER CODE BEGIN 1 */

  /* USER CODE END 1 */

  // MCU Configuration

  // Reset peripherals, initialize the flash interface and Systick.
  HAL_Init();
```

```c
    /* USER CODE BEGIN Init */

    /* USER CODE END Init */

    /* Configure the system clock */
    SystemClock_Config();

    /* USER CODE BEGIN SysInit */

    /* USER CODE END SysInit */

    /* Initialize all configured peripherals */
    MX_GPIO_Init();
    MX_TIM2_Init();

    /* USER CODE BEGIN 2 */

    /* USER CODE END 2 */

    /* Infinite loop */
    /* USER CODE BEGIN WHILE */
    while (1)
    {
    /* USER CODE END WHILE */

    /* USER CODE BEGIN 3 */
        int timerValue = __HAL_TIM_GET_COUNTER(&htim2);
        if(timerValue == 150)
          HAL_GPIO_WritePin(LD2_GPIO_Port, LD2_Pin, GPIO_PIN_SET);
        if(timerValue == 399)
          HAL_GPIO_WritePin(LD2_GPIO_Port, LD2_Pin, GPIO_PIN_RESET);

    }
    /* USER CODE END 3 */

}

/** System Clock Configuration
*/
void SystemClock_Config(void)
{

  RCC_OscInitTypeDef RCC_OscInitStruct;
  RCC_ClkInitTypeDef RCC_ClkInitStruct;

  // Initialize the CPU, AHB and APB buss clocks
  RCC_OscInitStruct.OscillatorType = RCC_OSCILLATORTYPE_HSI;
  RCC_OscInitStruct.HSIState = RCC_HSI_ON;
  RCC_OscInitStruct.HSICalibrationValue = 16;
  RCC_OscInitStruct.PLL.PLLState = RCC_PLL_ON;
  RCC_OscInitStruct.PLL.PLLSource = RCC_PLLSOURCE_HSI;
  RCC_OscInitStruct.PLL.PLLMUL = RCC_PLL_MUL16;
```

```
if (HAL_RCC_OscConfig(&RCC_OscInitStruct) != HAL_OK)
{
  _Error_Handler(__FILE__, __LINE__);
}

// Initialize the CPU, AHB and APB buss clocks

RCC_ClkInitStruct.ClockType = RCC_CLOCKTYPE_HCLK|RCC_CLOCKTYPE_SYSCLK
                              |RCC_CLOCKTYPE_PCLK1|RCC_CLOCKTYPE_PCLK2;
RCC_ClkInitStruct.SYSCLKSource = RCC_SYSCLKSOURCE_PLLCLK;
RCC_ClkInitStruct.AHBCLKDivider = RCC_SYSCLK_DIV1;
RCC_ClkInitStruct.APB1CLKDivider = RCC_HCLK_DIV2;
RCC_ClkInitStruct.APB2CLKDivider = RCC_HCLK_DIV1;

if (HAL_RCC_ClockConfig(&RCC_ClkInitStruct, FLASH_LATENCY_2) != HAL_OK)
{
  _Error_Handler(__FILE__, __LINE__);
}

  /**Configure the Systick interrupt time
  */
HAL_SYSTICK_Config(HAL_RCC_GetHCLKFreq()/1000);

  /**Configure the Systick
  */
HAL_SYSTICK_CLKSourceConfig(SYSTICK_CLKSOURCE_HCLK);

/* SysTick_IRQn interrupt configuration */
HAL_NVIC_SetPriority(SysTick_IRQn, 0, 0);
}

/* TIM2 init function */
static void MX_TIM2_Init(void)
{
  TIM_ClockConfigTypeDef sClockSourceConfig;
  TIM_MasterConfigTypeDef sMasterConfig;

  htim2.Instance = TIM2;
  htim2.Init.Prescaler = 63999; // Establishes a 1 KHz clock input
  htim2.Init.CounterMode = TIM_COUNTERMODE_UP;
  htim2.Init.Period = 500; // Establishes the 0.5 second period
  htim2.Init.ClockDivision = TIM_CLOCKDIVISION_DIV1;
  htim2.Init.AutoReloadPreload = TIM_AUTORELOAD_PRELOAD_DISABLE;
  if (HAL_TIM_Base_Init(&htim2) != HAL_OK)
  {
    _Error_Handler(__FILE__, __LINE__);
  }

  sClockSourceConfig.ClockSource = TIM_CLOCKSOURCE_INTERNAL;
  if (HAL_TIM_ConfigClockSource(&htim2, &sClockSourceConfig) != HAL_OK)
  {
    _Error_Handler(__FILE__, __LINE__);
```

```
    }

    sMasterConfig.MasterOutputTrigger = TIM_TRGO_RESET;
    sMasterConfig.MasterSlaveMode = TIM_MASTERSLAVEMODE_DISABLE;
    if (HAL_TIMEx_MasterConfigSynchronization(&htim2, &sMasterConfig) != HAL_OK)
    {
      _Error_Handler(__FILE__, __LINE__);
    }

    HAL_TIM_Base_Start(&htim2);   // This starts the TIM2 timer
}

/** Configure pins as
        * Analog
        * Input
        * Output
        * EVENT_OUT
        * EXTI
     PA2    ------> USART2_TX
     PA3    ------> USART2_RX
*/
static void MX_GPIO_Init(void)
{
  GPIO_InitTypeDef GPIO_InitStruct;

  /* GPIO Ports Clock Enable */
  __HAL_RCC_GPIOC_CLK_ENABLE();
  __HAL_RCC_GPIOF_CLK_ENABLE();
  __HAL_RCC_GPIOA_CLK_ENABLE();
  __HAL_RCC_GPIOB_CLK_ENABLE();

  /*Configure GPIO pin Output Level */
  HAL_GPIO_WritePin(LD2_GPIO_Port, LD2_Pin, GPIO_PIN_RESET);

  /*Configure GPIO pin : B1_Pin */
  GPIO_InitStruct.Pin = B1_Pin;
  GPIO_InitStruct.Mode = GPIO_MODE_IT_FALLING;
  GPIO_InitStruct.Pull = GPIO_NOPULL;
  HAL_GPIO_Init(B1_GPIO_Port, &GPIO_InitStruct);

  /*Configure GPIO pins : USART_TX_Pin USART_RX_Pin */
  GPIO_InitStruct.Pin = USART_TX_Pin|USART_RX_Pin;
  GPIO_InitStruct.Mode = GPIO_MODE_AF_PP;
  GPIO_InitStruct.Pull = GPIO_NOPULL;
  GPIO_InitStruct.Speed = GPIO_SPEED_FREQ_LOW;
  GPIO_InitStruct.Alternate = GPIO_AF7_USART2;
  HAL_GPIO_Init(GPIOA, &GPIO_InitStruct);

  /*Configure GPIO pin : LD2_Pin */
  GPIO_InitStruct.Pin = LD2_Pin;
  GPIO_InitStruct.Mode = GPIO_MODE_OUTPUT_PP;
  GPIO_InitStruct.Pull = GPIO_NOPULL;
```

```
  GPIO_InitStruct.Speed = GPIO_SPEED_FREQ_LOW;
  HAL_GPIO_Init(LD2_GPIO_Port, &GPIO_InitStruct);

}

/* USER CODE BEGIN 4 */

/* USER CODE END 4 */

/**
  * @brief  This function is executed in case of error occurrence.
  * @param  None
  * @retval None
  */
void _Error_Handler(char * file, int line)
{
  /* USER CODE BEGIN Error_Handler_Debug */
  /* User can add his own implementation to report the HAL error return state */
  while(1)
  {
  }
  /* USER CODE END Error_Handler_Debug */
}

#ifdef USE_FULL_ASSERT

/**
  * @brief Reports the name of the source file and the source line number
  * where the assert_param error has occurred.
  * @param file: pointer to the source file name
  * @param line: assert_param error line source number
  * @retval None
  */
void assert_failed(uint8_t* file, uint32_t line)
{
  /* USER CODE BEGIN 6 */
  /* User can add his own implementation to report the file name and line number,
  ex: printf("Wrong parameters value: file %s on line %d\r\n", file, line) */
  /* USER CODE END 6 */

}

#endif

/**
  * @}
  */

/**
  * @}
*/

/***** (C) COPYRIGHT STMicroelectronics *****/
```

Most of this code is a repeat of the main.c code shown in the interrupt demonstration project discussed in the previous chapter. The key differences are the addition of a MX_GPIO_Init method and some new operational code required in the forever loop portion of the main method.

The MX_GPIO_Init method configures the TIM_HandleTypeDef, a typed C struct with specific values to have the timer reset every 0.5 s. The struct is named htim2 and is both initialized and started in this method.

The forever loop in the main method has code, which continually checks the timer count and will turn on the LED when the count reaches 150. It will remain on until the count progresses to a value of 399, at which point the LED will be reset. These statements result in the LED being on for 0.25 s and off for 0.25 s.

Test Run

I built the project and uploaded it into the Nucleo-64 board. The user LED immediately began to blink at a 0.5-s rate, which confirmed the program worked as desired. The next step in this timer demonstration series is to implement an interrupt-driven version that will also blink the LED, but not require any specialized code in the main method's forever loop.

Interrupt-Driven Blink LED Timer Demonstration

This demonstration will also blink an LED. In this demonstration, I will use an interrupt approach, thus eliminating any specialized code in the forever loop main that continually reads the timer counter value and reacts when certain values are reached.

You will need to generate a new project using the CubeMX application. I named this project Timer_int to reflect its purpose. You will also need to select the timer clock source as has been done in the previous demonstration. Simply use the internal high-speed clock as the source, which will trigger the application to include all the TIM2 timer initialization and configuration code. However, this time you will also have to set up the timer to trigger an interrupt when its period is elapsed. This is easily done using the CubeMX Configuration tab. You should see timer TIM2 appear in the tab, but first ensure you have selected the clock source. Otherwise timer TIM2 will not appear. Click on the TIM2 icon and then

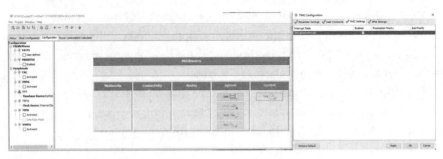

Figure 7-2 *Configuring timer TIM2 interrupt functionality.*

click on the enable global interrupt check box. The Configuration tab and the enable global interrupt check box are both shown in Figure 7-2.

I will not repeat the extensive code listing previously shown, but will instead just present the code segments needed to implement the interrupt-based code.

The first part is the easy one, where the forever loop is changed to a null action:

```
while(1);
```

The only other change that needs to be done is to set the timer period to 250 to ensure the LED blinks twice a second. The timer will now interrupt at the appropriate interval without any special code needed to monitor real-time counter values.

The following three statements must be added just before the forever loop in the `main` method:

```
// Start the TIM2 timer in the interrupt mode
HAL_TIM_Base_Start_IT(&htim2);
// Set the interrupt priority
HAL_NVIC_SetPriority(TIM2_IRQn, 0, 0);
// Enable the peripheral IRQ
HAL_NVIC_EnableIRQ(TIM2_IRQn);
```

The TIM2 IRQ handler method must be defined in the main.c file as follows:

```
void TIM2_IRQHandler(void) {
    HAL_TIM_IRQHandler(&htim2);
}
```

Finally, a callback method must also be defined, which toggles the LED pin every time the TIM2 timer generates an interrupt:

```
// This callback is automatically called by the HAL on the TIM2 event
```

```
void HAL_TIM_PeriodElapsedCallback(*htim2) {
    if(htim2->Instance == TIM2)
    HAL_GPIO_TogglePin(LD2_GPIO_Port, LD2_Pin);
}
```

Test Run

I made the modifications to the main.c file, built the project and then uploaded it into the Nucleo-64 board. The user LED immediately began to blink at a 0.5 second rate, which confirmed the program worked as desired.

Multi-rate Interrupt-Driven Blink LED Timer Demonstration

This demonstration is an expansion of the previous demonstration in which the user LED will be repeatedly blinked at two different rates, one at 0.5 s and the other at a 1.0-s rate. A timer interrupt will also be used in this program but with a significant difference between the single- and multi-rate implementations. In the single-rate program, the LED is directly controlled in the callback function. In the multi-rate version, the callback function simply increments a counter variable, which is then used in the forever loop to control the LED. The incrementing counter variable `timCnt` causes the LED to be set or reset depending on its value. The following listing is the complete main.c listing. I have also added additional comments regarding the forever loop code snippet following the listing.

```
// STM disclaimer goes here
// Includes
#include "main.h"
#include "stm32f3xx_hal.h"

/* USER CODE BEGIN Includes */

/* USER CODE END Includes */

// Private variables
TIM_HandleTypeDef htim2;

/* USER CODE BEGIN PV */
// Private variables
int timCnt = -1;
/* USER CODE END PV */
```

```
// Private function prototypes
void SystemClock_Config(void);
static void MX_GPIO_Init(void);
static void MX_TIM2_Init(void);

/* USER CODE BEGIN PFP */
// Private function prototypes

/* USER CODE END PFP */

/* USER CODE BEGIN 0 */

/* USER CODE END 0 */

int main(void)
{

  /* USER CODE BEGIN 1 */

  /* USER CODE END 1 */

  // MCU Configuration

  // Reset peripherals, initialize the flash interface and Systick
  HAL_Init();

  /* USER CODE BEGIN Init */

  /* USER CODE END Init */

  /* Configure the system clock */
  SystemClock_Config();

  /* USER CODE BEGIN SysInit */

  /* USER CODE END SysInit */

  /* Initialize all configured peripherals */
  MX_GPIO_Init();
  MX_TIM2_Init();

  /* USER CODE BEGIN 2 */
  HAL_TIM_Base_Start_IT(&htim2);
  /* USER CODE END 2 */

  /* Infinite loop */
  /* USER CODE BEGIN WHILE */
  while (1)
  {
  /* USER CODE END WHILE */

    /* USER CODE BEGIN 3 */
```

```
        if(timCnt == 0) // start the 0.5 second intervals
            {
              HAL_GPIO_WritePin(LD2_GPIO_Port, LD2_Pin, GPIO_PIN_SET);
            }
        if(timCnt == 2)
            {
              HAL_GPIO_WritePin(LD2_GPIO_Port, LD2_Pin, GPIO_PIN_RESET);
        }
        if( timCnt == 4)
            {
        HAL_GPIO_WritePin(LD2_GPIO_Port, LD2_Pin, GPIO_PIN_SET);
            }
        if(timCnt == 6)
            {
              HAL_GPIO_WritePin(LD2_GPIO_Port, LD2_Pin, GPIO_PIN_RESET);
        }
        if(timCnt == 8)
            {
              HAL_GPIO_WritePin(LD2_GPIO_Port, LD2_Pin, GPIO_PIN_SET);
            }
        if(timCnt == 10)
            {
              HAL_GPIO_WritePin(LD2_GPIO_Port, LD2_Pin, GPIO_PIN_RESET);
        }
        if(timCnt == 12)
            {
              HAL_GPIO_WritePin(LD2_GPIO_Port, LD2_Pin, GPIO_PIN_SET);
            }
        if(timCnt == 14)
            {
              HAL_GPIO_WritePin(LD2_GPIO_Port, LD2_Pin, GPIO_PIN_RESET);
        }
        if( timCnt == 16) // start the 1.0 second intervals
            {
              HAL_GPIO_WritePin(LD2_GPIO_Port, LD2_Pin, GPIO_PIN_SET);
             }
        if(timCnt == 20)
            {
              HAL_GPIO_WritePin(LD2_GPIO_Port, LD2_Pin, GPIO_PIN_RESET);
        }
        if( timCnt == 24)
            {
              HAL_GPIO_WritePin(LD2_GPIO_Port, LD2_Pin, GPIO_PIN_SET);
            }
        if(timCnt == 28)
            {
              HAL_GPIO_WritePin(LD2_GPIO_Port, LD2_Pin, GPIO_PIN_RESET);
            }
  /* USER CODE END 3 */
  } // end of while loop
} // end of main method
```

```
// System Clock Configuration
void SystemClock_Config(void)
{
  RCC_OscInitTypeDef RCC_OscInitStruct;
  RCC_ClkInitTypeDef RCC_ClkInitStruct;

  // Initializes the CPU, AHB and APB buss clocks
  RCC_OscInitStruct.OscillatorType = RCC_OSCILLATORTYPE_HSI;
  RCC_OscInitStruct.HSIState = RCC_HSI_ON;
  RCC_OscInitStruct.HSICalibrationValue = 16;
  RCC_OscInitStruct.PLL.PLLState = RCC_PLL_ON;
  RCC_OscInitStruct.PLL.PLLSource = RCC_PLLSOURCE_HSI;
  RCC_OscInitStruct.PLL.PLLMUL = RCC_PLL_MUL16;
  if (HAL_RCC_OscConfig(&RCC_OscInitStruct) != HAL_OK)
  {
    _Error_Handler(__FILE__, __LINE__);
  }

  // Initializes the CPU, AHB and APB buss clocks
  RCC_ClkInitStruct.ClockType = RCC_CLOCKTYPE_HCLK|RCC_CLOCKTYPE_SYSCLK
                              |RCC_CLOCKTYPE_PCLK1|RCC_CLOCKTYPE_PCLK2;
  RCC_ClkInitStruct.SYSCLKSource = RCC_SYSCLKSOURCE_PLLCLK;
  RCC_ClkInitStruct.AHBCLKDivider = RCC_SYSCLK_DIV1;
  RCC_ClkInitStruct.APB1CLKDivider = RCC_HCLK_DIV2;
  RCC_ClkInitStruct.APB2CLKDivider = RCC_HCLK_DIV1;

  if (HAL_RCC_ClockConfig(&RCC_ClkInitStruct, FLASH_LATENCY_2) != HAL_OK)
  {
    _Error_Handler(__FILE__, __LINE__);
  }

  // Configure the Systick interrupt time
  HAL_SYSTICK_Config(HAL_RCC_GetHCLKFreq()/1000);

  // Configure the Systick
  HAL_SYSTICK_CLKSourceConfig(SYSTICK_CLKSOURCE_HCLK);

  /* SysTick_IRQn interrupt configuration */
  HAL_NVIC_SetPriority(SysTick_IRQn, 0, 0);
}

/* TIM2 init function */
static void MX_TIM2_Init(void)
{
  TIM_ClockConfigTypeDef sClockSourceConfig;
  TIM_MasterConfigTypeDef sMasterConfig;

  htim2.Instance = TIM2;
  htim2.Init.Prescaler = 63999;
  htim2.Init.CounterMode = TIM_COUNTERMODE_UP;
  htim2.Init.Period = 250; // now a 0.25 second period
  htim2.Init.ClockDivision = TIM_CLOCKDIVISION_DIV1;
```

```
  htim2.Init.AutoReloadPreload = TIM_AUTORELOAD_PRELOAD_DISABLE;
  if (HAL_TIM_Base_Init(&htim2) != HAL_OK)
  {
    _Error_Handler(__FILE__, __LINE__);
  }

  sClockSourceConfig.ClockSource = TIM_CLOCKSOURCE_INTERNAL;
  if (HAL_TIM_ConfigClockSource(&htim2, &sClockSourceConfig) != HAL_OK)
  {
    _Error_Handler(__FILE__, __LINE__);
  }

  sMasterConfig.MasterOutputTrigger = TIM_TRGO_RESET;
  sMasterConfig.MasterSlaveMode = TIM_MASTERSLAVEMODE_DISABLE;
  if (HAL_TIMEx_MasterConfigSynchronization(&htim2, &sMasterConfig) != HAL_OK)
  {
    _Error_Handler(__FILE__, __LINE__);
  }

}

void HAL_TIM_PeriodElapsedCallback(TIM_HandleTypeDef *htim)
{
        if(htim -> Instance == TIM2)
        {
          timCnt++;
            if(timCnt == 32) // 8 seconds overall
            {
                timCnt = -1;
            }
        }
}

/** Configure pins as
        * Analog
        * Input
        * Output
        * EVENT_OUT
        * EXTI
     PA2      ------> USART2_TX
     PA3      ------> USART2_RX
*/
static void MX_GPIO_Init(void)
{

  GPIO_InitTypeDef GPIO_InitStruct;

  /* GPIO Ports Clock Enable */
  __HAL_RCC_GPIOC_CLK_ENABLE();
  __HAL_RCC_GPIOF_CLK_ENABLE();
  __HAL_RCC_GPIOA_CLK_ENABLE();
  __HAL_RCC_GPIOB_CLK_ENABLE();
```

```
    /*Configure GPIO pin Output Level */
    HAL_GPIO_WritePin(LD2_GPIO_Port, LD2_Pin, GPIO_PIN_RESET);

    /*Configure GPIO pin : B1_Pin */
    GPIO_InitStruct.Pin = B1_Pin;
    GPIO_InitStruct.Mode = GPIO_MODE_IT_FALLING;
    GPIO_InitStruct.Pull = GPIO_NOPULL;
    HAL_GPIO_Init(B1_GPIO_Port, &GPIO_InitStruct);

    /*Configure GPIO pins : USART_TX_Pin USART_RX_Pin */
    GPIO_InitStruct.Pin = USART_TX_Pin|USART_RX_Pin;
    GPIO_InitStruct.Mode = GPIO_MODE_AF_PP;
    GPIO_InitStruct.Pull = GPIO_NOPULL;
    GPIO_InitStruct.Speed = GPIO_SPEED_FREQ_LOW;
    GPIO_InitStruct.Alternate = GPIO_AF7_USART2;
    HAL_GPIO_Init(GPIOA, &GPIO_InitStruct);

    /*Configure GPIO pin : LD2_Pin */
    GPIO_InitStruct.Pin = LD2_Pin;
    GPIO_InitStruct.Mode = GPIO_MODE_OUTPUT_PP;
    GPIO_InitStruct.Pull = GPIO_NOPULL;
    GPIO_InitStruct.Speed = GPIO_SPEED_FREQ_LOW;
    HAL_GPIO_Init(LD2_GPIO_Port, &GPIO_InitStruct);

}

/* USER CODE BEGIN 4 */

/* USER CODE END 4 */

/**
  * @brief  This function is executed in case of error occurrence.
  * @param  None
  * @retval None
  */
void _Error_Handler(char * file, int line)
{
  /* USER CODE BEGIN Error_Handler_Debug */
  /* User can add his own implementation to report the HAL error return state */
  while(1)
  {
  }
  /* USER CODE END Error_Handler_Debug */
}

#ifdef USE_FULL_ASSERT

/**
  * @brief Reports the name of the source file and the source line number
  * where the assert_param error has occurred.
  * @param file: pointer to the source file name
  * @param line: assert_param error line source number
```

```
 * @retval None
 */
void assert_failed(uint8_t* file, uint32_t line)
{
  /* USER CODE BEGIN 6 */
  /* User can add his own implementation to report the file name and line number,
     ex: printf("Wrong parameters value: file %s on line %d\r\n", file, line) */
  /* USER CODE END 6 */

}

#endif

/**
  * @}
  */

/**
  * @}
  */

/**** (C) COPYRIGHT STMicroelectronics ****/
```

The forever loop contains a series of if statements that will be triggered based on specific counter values. The statements within the if statements either set or reset the user LED. Each counter increment takes 0.25 s to complete, which means that two increments will take 0.5 s. The LED will toggle at two times per second for the first 4 s. The second set of 4 s will toggle the LED once every second.

This code format is rather lengthy, but it does trade code space for simplicity. It is rather easy to follow the algorithm step-by-step to determine how the LED is controlled using this format.

```
   while (1)
  {
 /* USER CODE END WHILE */

 /* USER CODE BEGIN 3 */
      if(timCnt == 0) // start the 0.5 second intervals
        {
          HAL_GPIO_WritePin(LD2_GPIO_Port, LD2_Pin, GPIO_PIN_SET);
        }
      if(timCnt == 2)
        {
          HAL_GPIO_WritePin(LD2_GPIO_Port, LD2_Pin, GPIO_PIN_RESET);
    }
      if( timCnt == 4)
        {
```

```
        HAL_GPIO_WritePin(LD2_GPIO_Port, LD2_Pin, GPIO_PIN_SET);
        }
      if(timCnt == 6)
      {
        HAL_GPIO_WritePin(LD2_GPIO_Port, LD2_Pin, GPIO_PIN_RESET);
    }
      if(timCnt == 8)
      {
        HAL_GPIO_WritePin(LD2_GPIO_Port, LD2_Pin, GPIO_PIN_SET);
      }
      if(timCnt == 10)
      {
          HAL_GPIO_WritePin(LD2_GPIO_Port, LD2_Pin, GPIO_PIN_RESET);
    }
      if(timCnt == 12)
      {
    HAL_GPIO_WritePin(LD2_GPIO_Port, LD2_Pin, GPIO_PIN_SET);
      }
      if(timCnt == 14)
      {
          HAL_GPIO_WritePin(LD2_GPIO_Port, LD2_Pin, GPIO_PIN_RESET);
    }
      if( timCnt == 16) // start the 1.0 second intervals
      {
    HAL_GPIO_WritePin(LD2_GPIO_Port, LD2_Pin, GPIO_PIN_SET);
      }
      if(timCnt == 20)
      {
          HAL_GPIO_WritePin(LD2_GPIO_Port, LD2_Pin, GPIO_PIN_RESET);
    }
      if( timCnt == 24)
      {
          HAL_GPIO_WritePin(LD2_GPIO_Port, LD2_Pin, GPIO_PIN_SET);
      }
      if(timCnt == 28)
      {
          HAL_GPIO_WritePin(LD2_GPIO_Port, LD2_Pin, GPIO_PIN_RESET);
    }

  /* USER CODE END 3 */
  }
```

Test Run

I built the project and then uploaded it into the Nucleo-64 board. The user LED immediately began to blink at a 0.5-s rate for 4 s and then at a 1.0-s rate for 4 s, which confirmed that the program worked as desired.

Modification to the Multi-rate Program

I changed code snippet in the above forever code to make that much more compact. That modification is listed below.

```
while (1)
{
/* USER CODE END WHILE */

/* USER CODE BEGIN 3 */
   if((timCnt % 2 == 0) && (timCnt < 16))
       {
              HAL_GPIO_TogglePin(LD2_GPIO_Port, LD2_Pin);
       }
       if((timCnt % 4 == 0) && (timCnt >= 16))
       {
              HAL_GPIO_TogglePin(LD2_GPIO_Port, LD2_Pin);
       }
}
   /* USER CODE END 3 */
}
```

The program functioned just about the same with this modified code, which used the modulus operator along with a toggle method to control the LED. The code is much smaller than the original, but it is more complex and harder to understand and consequently maintain. This change highlights an interesting challenge that developers constantly face. This challenge is whether or not to use simpler code, which can take more code space versus using compact and "efficient" code, which is likely harder to understand and modify.

Test Run

I built the modified project and then uploaded it into the Nucleo-64 board. The user LED immediately began to blink at a 0.5-s rate for 4 s and then at a 1.0-s rate for 4 s, which confirmed the modified program worked the same as the original program. I did notice that the LED seemed to flicker more using this version vice the original. I can only attribute the flicker to the use of the toggle method versus the firm set and reset methods.

This last demonstration completes all discussions and demonstrations that I wished to convey regarding interrupts and timers. This chapter really is just the tip of the proverbial iceberg because there is a tremendous amount of information on these topics that I could not cover. There are almost 2,000 pages alone in STM's HAL user manual. As always, the STM datasheets, and user and reference manuals will be an invaluable resource to learn about the many additional features and functions that you can carry out with interrupts and timers.

Summary

I started the chapter discussions with the focus on timers, which are important peripherals because they alleviate the MCU from having to directly perform timing functions. I explained the five available categories of STM timers, of which there are three types provided on the project MCU. These are basic, general purpose (GP), and advanced.

I next showed you how to configure a timer to perform a specific timing event. This discussion was followed by a demonstration in which the onboard LED blinked once per second. This demonstration did not use an interrupt, which was the focus of the second timer demonstration.

In the second timer demonstration, I showed how easy it was to integrate a timing operation with an interrupt process. In this demonstration also the onboard LED blinked, but it blinked using only the interrupt process.

The third timer demonstration showed how to restructure the interrupt process such that a multi-rate LED blink could be implemented. The interrupt callback method controlled a counter variable, which in turn controlled the actual LED activation within the forever loop.

In the final timer demonstration, I changed the original forever loop code with a much more compact and complex version with the same functionality. I did this to illustrate how developers must constantly decide on how to write code, simpler and consuming more code space or very compact and complex, which is harder to understand.

8

Bit Serial Communications

This chapter covers how a MCU uses bit serial communication to bidirectionally transfer data between it and external devices and systems. The primary peripheral devices used for this function are the universal asynchronous receiver transmitter (UART) and the universal synchronous/asynchronous receiver/transmitter (USART). The principal difference between these peripherals is that the USART can use a synchronizing clock pulse train between nodes, while the UART is completely self-synchronizing. These differences are explained in more detail in the following section "UARTs and USARTs."

UARTs and USARTs

Figure 8-1 is a block diagram showing two UARTs connected, where Node A is transmitting to Node B.

The digital pulse train between the nodes is composed of one start bit, eight data bits, and one stop bit. These bits all together are considered to be a single frame. You should note that the start and stop bits are 50% longer in duration then

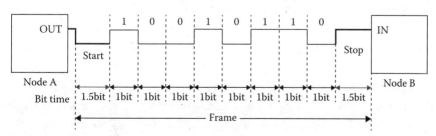

Figure 8-1 *UART block diagram.*

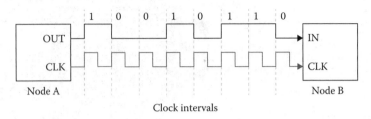

Figure 8-2 *USART block diagram.*

the data bits. This difference in pulse duration is how the UART distinguishes the start and stop bits from the data bits. The receiving node must detect the start bit in order to determine when the frame starts. The receiving node also has been configured to accept a predetermined number of data bits and whether or not there will be one or two stop bits. UARTs can be set to send seven or eight data bits as well as one or two stop bits. There is also a provision for sending an additional parity bit, which aids in detecting transmission errors. Often, parity bits are not used, especially in cases where the transmission channel is not noisy or transmission line delays are not present.

The USART is a variant of the UART because an additional clock line can be used between nodes, as shown in Figure 8-2.

There is no need for start and stop bits to be used when two USART nodes are interconnected, as shown in Figure 8-2. The data bits are clearly synchronized to the clock pulses, which shortens the frame length and improves channel communication efficiency. However, seven or eight data bits may be set as well as a parity bit, as was the case for the UART node. It should also be obvious from the name that a USART can operate as a UART, if so desired.

Figure 8-3 is a block diagram that shows full-duplex communication setups using UART and USART nodes.

Full-duplex communications means that data can simultaneously be sent and received between nodes. This is strictly possible because each node's UART or USART is controlled by its own MCU.

Often, it is important to assess the state of a given node that it is able to actively participate in a communications link. It can happen that a receiving node's MCU is temporarily blocked for some reason and while a remote node's UART can send data, the receiving node may not be able to process it. For this reason, there are protocol control lines that are provided in UART and USART hardware to ensure that data is sent and received when allowed. The two primary control lines are

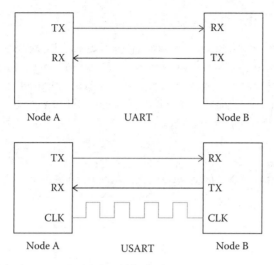

Figure 8-3 *Full-duplex communications block diagram.*

request-to-send (RTS) and clear-to-send (CTS). A node that desires to transmit data would first raise or set the RTS line. The receiving node would then raise or set the CTS line indicating that it is okay to send the data. This process is also known as request/acknowledge and is commonly used in full-duplex communications links. However, this protocol is really not required for the book projects because the project board MCU will never be blocked and the receiving node will typically be a PC, which will also never be blocked.

There are two USARTs in the STM32F302R8 MCU use in the project board. These are logically referred to as USART1 and USART2. They are the primary means by which the MCU can send and receive data other than by processing data bits through the GPIO ports or by using one of the other communication protocols, such as the I2C and SPI bit-serial interfaces. The latter two are used mainly for sensor data, although there is a very nice I2C peripheral that can display text messages. Later in the chapter, I will discuss how to set up the USART2 USART to communicate with a PC terminal program. However, I now need to discuss how to initially configure a USART.

NOTE   *The software to initialize and configure UARTs and USARTs is identical. Do not worry that a USART is being discussed, while the function or typedef has UART in the name.*

USART Configuration

STM USARTs are configured in a similar manner as GPIO pins, interrupts, and timers that I have previously discussed. STM USARTs are initialized using a C struct named `UART_HandleTypeDef`. It is configured using a C struct named `UART_InitTypeDef`. I will first discuss the `UART_HandleTypeDef`, which contains the following members:

```
USART_TypeDef                   *Instance;
UART_InitTypeDef                Init;
UART_AdvFeatureInitTypeDef      AdvancedInit;
uint8_t                         *pTxBufPtr;
uint16_t                        TxXferSize;
uint16_t                        TxXferCount;
uint8_t                         *pRxBuffPtr;
uint16_t                        RxXferSize;
uint16_t                        RxXferCount;
DMA_HandleTypeDef               *hdmatx;
DMA_HandleTypeDef               *hdmarx;
HAL_LockTypeDef                 Lock;
__IO HAL_UART_StateTypeDef      State;
__IO HAL_UART_ErrorTypeDef      ErrorCode;
```

Some of the more important members of this struct are discussed below:

- Instance: This is the pointer to the USART descriptor. For example, USART2 is the descriptor of the UART associated to the ST-LINK interface.

- Init: This is an instance of the C struct UART_InitTypeDef, which is used to configure the UART interface.

- AdvancedInit: Used to configure more advanced UART features like the automatic BaudRate detection and TX/RX pin swaps.

- pTxBuffPtr: Points to the transmit buffer used as the data source for transmitted bytes.

- TxXferCount: Holds the count of transmitted bytes.

- pRxBuffPtr: Points to the receive buffer where incoming bytes will be stored.

- RxXferCount: Holds the count of received bytes.

- Lock: This field is used internally by the HAL to lock concurrent accesses to UART interfaces.

The C struct `UART_InitTypeDef` contains the following members:

```
uint32_t BaudRate;
uint32_t WordLength;
uint32_t StopBits;
uint32_t Parity;
uint32_t Mode;
uint32_t HwFlowCtl;
uint32_t OverSampling;
```

These struct members represent the following USART parameters and settings:

Baudrate: This parameter is the connection speed in bits per second. Table 8-1 shows a chart of standard and nonstandard baud rates that are available with the STM USART peripheral.

- WordLength: This value specifies the number of data bits transmitted or received in a frame. This field can assume one of the following defines:

  - UART_WORDLENGTH_8B

  - UART_WORDLENGTH_9B

- StopBits: This field specifies the number of stop bits transmitted. This field can assume one of the following defines:

  - UART_STOPBITS_1

  - UART_STOPBITS_2

Baud Rate Chart		
S Number	**Baud**	**Remark**
2	2400	Standard
3	9600	Standard
4	19200	Standard
5	38400	Standard
6	57600	Standard
7	115200	Standard
8	230400	Non-standard, not recommended
9	460800	Non-standard, not recommended
10	921600	Non-standard, not recommended
11	2000000	Non-standard, not recommended
12	3000000	Non-standard, not recommended

Table 8-1 *Baud Rate Chart*

- Parity: This value indicates the parity mode. This field can assume one of the following defines:

  - UART_PARITY_NONE

  - UART_PARITY_EVEN

  - UART_PARITY_ODD

- Mode: This value specifies whether the RX and/or TX mode is enabled or disabled. This field can assume one of the following defines:

  - UART_MODE_RX

  - UART_MODE_TX

  - UART_MODE_TX_RX

- HwFlowCtl :This value specifies whether the RS2326 Hardware Flow Control mode is enabled or disabled. This field can assume one of the following defines:

  - UART_HWCONTROL_NONE

  - UART_HWCONTROL_RTS

  - UART_HWCONTROL_CTS

  - UART_HWCONTROL_RTS_CTS

OverSampling: This value helps discriminate noise from true signal levels. The selected oversampling rate is a tradeoff between noise immunity and maximum possible communication speed. This field can assume one of the following defines:

- UART_OVERSAMPLING_8

- UART_OVERSAMPLING_16

The above discussion details how a USART can be initialized and configured, but I still need to explain how to set up a communication link between the project board and a remote node, which will be the PC connected to the board.

Windows Terminal Program

You will need a terminal program installed in order to display the text messages being transmitted from the demonstration board. There are a number of terminal programs that are free and readily available to download. I chose a program named Realterm that is available from https://sourceforge.net/projects/realterm/.

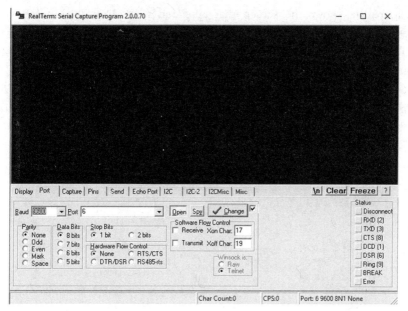

Figure 8-4 *Realterm Port screen.*

It is full-featured terminal program that I found very easy to configure and run. Figure 8-4 shows the Realterm screen after I clicked on the Port tab.

Note that I selected the following parameters to match those configured in the STM program:

- Baud—9600
- Parity— None
- Data Bits—8
- Stop Bits—1
- Hardware Flow Control—None
- Port—6

The last parameter named Port deserves some further explanation. This value refers to the virtual com port number that was created when I installed the ST-LINK Virtual COM Port USB driver into the host computer. In Windows 10, I had to use the following lengthy sequence just to find the port number:

Control Panel → Hardware and Sound → Devices and Printers → Device Manager → Ports (COM & LPT)

Figure 8-5 shows the screen that appeared when I clicked on the Ports (COM & LPT) selection.

The USART2 peripheral is connected to the ST-LINK circuit, as shown in the Figure 8-6 schematic.

GPIO pins PA2 and PA3 are connected as USART_TX and USART_RX, respectively. It is also possible to use these pins as an external USART by shorting

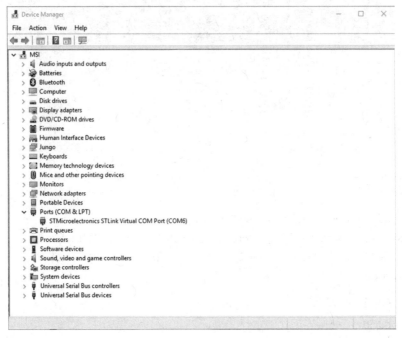

Figure 8-5 *Ports screen.*

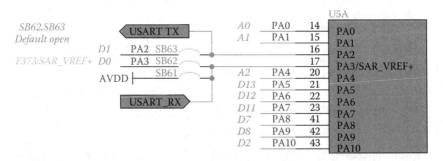

Figure 8-6 *USART2 connection to ST-LINK schematic.*

the solder bridges SB62 and SB63. The pins would then be available on both the Arduino and Morpho connectors, as discussed in a previous chapter.

The Realterm program will be all set to display the USART output when the program is run. However, there is an additional step that is required for the USART to work as desired. This is described in the next section.

Enabling USART2

Up until this program, I have not had to use the IP tree pane view in the CubeMX application to configure a peripheral. That has been done for me by the application because I was only using GPIO pins for a program and not any alternative functions. However, I will now need to enable the USART2 alternative function in order to have the communications link work. You configure USART2 when creating a new project, and the Pinout window is shown. Figure 8-7 shows how USART2 is configured by selecting Asynchronous from the Mode dropdown menu.

The CubeMX application will automatically insert an initialization method `static void MX_USART2_UART_Init(void)` in the main.c file when the USART2 function is enabled.

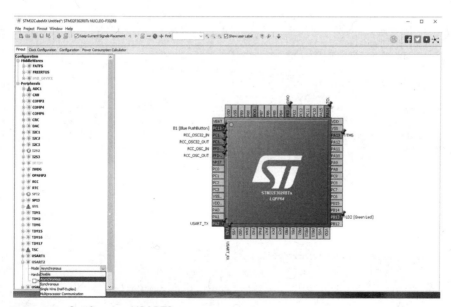

Figure 8-7 *Configuring USART2.*

You need to set the appropriate communications parameters in this method as shown in the following code snippet:

```
static void MX_USART2_UART_Init(void)
{
    huart2.Instance = USART2;
    huart2.Init.BaudRate = 9600;
    huart2.Init.WordLength = UART_WORDLENGTH_8B;
    huart2.Init.StopBits = UART_STOPBITS_1;
    huart2.Init.Parity = UART_PARITY_NONE;
    huart2.Init.Mode = UART_MODE_TX_RX;
    huart2.Init.HwFlowCtl = UART_HWCONTROL_NONE;
    huart2.Init.OverSampling = UART_OVERSAMPLING_16;
    huart2.Init.OneBitSampling = UART_ONE_BIT_SAMPLE_DISABLE;
    huart2.AdvancedInit.AdvFeatureInit = UART_ADVFEATURE_NO_INIT;
```

At this point, all the prerequisites have been introduced that are needed for an actual USART demonstration program.

USART Transmit Demonstration Program

The following program will display a continuously varying value in a terminal window. The purpose of this demonstration is to show how the USART can transmit data to a terminal program.

```
// STM disclaimer goes here
// Includes
#include "main.h"
#include "stm32f3xx_hal.h"

/* USER CODE BEGIN Includes */

/* USER CODE END Includes */

// Private variables
UART_HandleTypeDef huart2;

/* USER CODE BEGIN PV */
// Private variables
double value = 1.0;
char msg[20] = "";
/* USER CODE END PV */

// Private function prototypes
void SystemClock_Config(void);
static void MX_GPIO_Init(void);
static void MX_USART2_UART_Init(void);
```

```
/* USER CODE BEGIN PFP */
// Private function prototypes

/* USER CODE END PFP */

/* USER CODE BEGIN 0 */

/* USER CODE END 0 */

int main(void)
{

  /* USER CODE BEGIN 1 */

  /* USER CODE END 1 */

  // MCU Configuration

  // Reset all peripherals, initializes the flash and Systick
  HAL_Init();

  /* USER CODE BEGIN Init */

  /* USER CODE END Init */

  /* Configure the system clock */
  SystemClock_Config();

  /* USER CODE BEGIN SysInit */

  /* USER CODE END SysInit */

  /* Initialize all configured peripherals */
  MX_GPIO_Init();
  MX_USART2_UART_Init();

  /* USER CODE BEGIN 2 */

  /* USER CODE END 2 */

  /* Infinite loop */
  /* USER CODE BEGIN WHILE */
  while (1)
  {
  /* USER CODE END WHILE */

  /* USER CODE BEGIN 3 */

      sprintf(msg, "%f\r\n", value); // form the output string msg
      HAL_UART_Transmit(&huart2, (uint8_t*) msg, 10, 100);
      HAL_Delay(200);
```

```
    value = value + 0.01;
  }
  /* USER CODE END 3 */

}


// System Clock Configuration
void SystemClock_Config(void)
{

  RCC_OscInitTypeDef RCC_OscInitStruct;
  RCC_ClkInitTypeDef RCC_ClkInitStruct;

  // Initialize the CPU, AHB and APB buss clocks
  RCC_OscInitStruct.OscillatorType = RCC_OSCILLATORTYPE_HSI;
  RCC_OscInitStruct.HSIState = RCC_HSI_ON;
  RCC_OscInitStruct.HSICalibrationValue = 16;
  RCC_OscInitStruct.PLL.PLLState = RCC_PLL_ON;
  RCC_OscInitStruct.PLL.PLLSource = RCC_PLLSOURCE_HSI;
  RCC_OscInitStruct.PLL.PLLMUL = RCC_PLL_MUL16;
  if (HAL_RCC_OscConfig(&RCC_OscInitStruct) != HAL_OK)
  {
    _Error_Handler(__FILE__, __LINE__);
  }

    // Initializes the CPU, AHB and APB buss clocks
  RCC_ClkInitStruct.ClockType =
  RCC_CLOCKTYPE_HCLK|RCC_CLOCKTYPE_SYSCLK
  |RCC_CLOCKTYPE_PCLK1|RCC_CLOCKTYPE_PCLK2;
  RCC_ClkInitStruct.SYSCLKSource = RCC_SYSCLKSOURCE_PLLCLK;
  RCC_ClkInitStruct.AHBCLKDivider = RCC_SYSCLK_DIV1;
  RCC_ClkInitStruct.APB1CLKDivider = RCC_HCLK_DIV2;
  RCC_ClkInitStruct.APB2CLKDivider = RCC_HCLK_DIV1;

  if (HAL_RCC_ClockConfig(&RCC_ClkInitStruct, FLASH_LATENCY_2) !=
  HAL_OK)
  {
    _Error_Handler(__FILE__, __LINE__);
  }

  // Configure the Systick interrupt time
  HAL_SYSTICK_Config(HAL_RCC_GetHCLKFreq()/1000);

  // Configure the Systick
  HAL_SYSTICK_CLKSourceConfig(SYSTICK_CLKSOURCE_HCLK);

  /* SysTick_IRQn interrupt configuration */
  HAL_NVIC_SetPriority(SysTick_IRQn, 0, 0);
}
```

```
/* USART2 init function */
static void MX_USART2_UART_Init(void)
{
  huart2.Instance = USART2;
  huart2.Init.BaudRate = 9600;
  huart2.Init.WordLength = UART_WORDLENGTH_8B;
  huart2.Init.StopBits = UART_STOPBITS_1;
  huart2.Init.Parity = UART_PARITY_NONE;
  huart2.Init.Mode = UART_MODE_TX_RX;
  huart2.Init.HwFlowCtl = UART_HWCONTROL_NONE;
  huart2.Init.OverSampling = UART_OVERSAMPLING_16;
  huart2.Init.OneBitSampling = UART_ONE_BIT_SAMPLE_DISABLE;
  huart2.AdvancedInit.AdvFeatureInit = UART_ADVFEATURE_NO_INIT;
  if (HAL_UART_Init(&huart2) != HAL_OK)
  {
    _Error_Handler(__FILE__, __LINE__);
  }

}

/** Configure pins as
        * Analog
        * Input
        * Output
        * EVENT_OUT
        * EXTI
*/
static void MX_GPIO_Init(void)
{

  /* GPIO Ports Clock Enable */
  __HAL_RCC_GPIOF_CLK_ENABLE();
  __HAL_RCC_GPIOA_CLK_ENABLE();

}

/* USER CODE BEGIN 4 */

/* USER CODE END 4 */

/**
  * @brief  This function is executed in case of error occurrence.
  * @param  None
  * @retval None
  */
void _Error_Handler(char * file, int line)
{
  /* USER CODE BEGIN Error_Handler_Debug */
  /* User can add his own implementation to report the HAL error
return state */
  while(1)
```

```
    {
    }
    /* USER CODE END Error_Handler_Debug */
}

#ifdef USE_FULL_ASSERT

/**
    * @brief Reports the name of the source file and the source line number
    * where the assert_param error has occurred.
    * @param file: pointer to the source file name
    * @param line: assert_param error line source number
    * @retval None
    */
void assert_failed(uint8_t* file, uint32_t line)
{
    /* USER CODE BEGIN 6 */
    /* User can add his own implementation to report the file name and
line number,
        ex: printf("Wrong parameters value: file %s on line %d\r\n", file,
line) */
    /* USER CODE END 6 */

}

#endif

/**
    * @}
    */

/**
    * @}
*/

/**** (C) COPYRIGHT STMicroelectronics ****/
```

A small amount of functional code has been placed in the forever loop. The following code snippet shows this code with an explanation following the listing.

```
while(1) {
    sprintf(msg, "%f\r\n", value); // form the output
string msg
    HAL_UART_Transmit(&huart2, (uint8_t*) msg, 10, 100);
    HAL_Delay(200);
    value = value + 0.01;
}
```

The sprintf statement causes the double variable value to be converted to a string with a carriage return and line feed concatenated to the value. The next statement `HAL_UART_Transmit(&huart2, (uint8_t*) msg, 10, 100);` performs the actual transmission of the data from the MCU to the terminal program. The four arguments are as follow:

`&huart2`—the address of the USART2 object.

`(uint8_t*) msg`—the string to be displayed. It has been cast as uint8_t characters.

`10`—the maximum string size.

`100`—the maximum delay time in milliseconds.

The next statement `HAL_Delay(200);` introduces a short delay between displaying the value string. The last statement just slightly changes the `value` variable to help with the display.

Test Run

The program was then built and downloaded into the project board. I then started the Realterm terminal program to display the communication stream from the board. Figure 8-8 shows the terminal output for the UART transmit program.

This program will also be used as a template for other book projects that only need a terminal output.

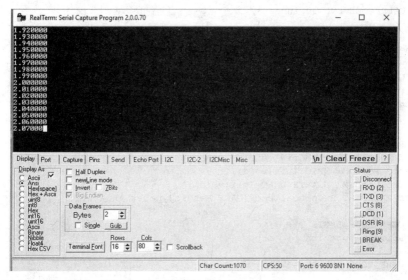

Figure 8-8   *Terminal output for the UART transmit program.*

USART Receive Demonstration Program

The following program will accept alphanumeric characters from a terminal program and will also display those characters. It will also detect a specific character sequence and light the user LED once it is detected. The purpose of this demonstration is to show how the USART can both receive and transmit data to/from a terminal program.

Because any received character will also trigger an interrupt, it is critical that the USART2 global interrupt should be enabled. This is done by clicking on the USART2 icon in the Configuration tab of the CubeMX program. Please follow the same procedure that was used in the timer interrupt example except you will be activating the USART2 global interrupt instead of the TIM2 timer interrupt. Enabling the USART2 interrupt will also cause the CubeMX application to automatically generate the associated callback function where all the required interrupt handling statements will be placed. All received characters are added to a data buffer within the callback function.

```
// STM disclaimer goes here
// Includes
#include "main.h"
#include "stm32f3xx_hal.h"

/* USER CODE BEGIN Includes */

/* USER CODE END Includes */

// Private variables
UART_HandleTypeDef huart2;

/* USER CODE BEGIN PV */
// Private variables
 char Rx_indx, Rx_data[2], Rx_Buffer[100], Transfer_cplt;
/* USER CODE END PV */

// Private function prototypes
void SystemClock_Config(void);
static void MX_GPIO_Init(void);
static void MX_USART2_UART_Init(void);
void HAL_UART_RxCpltCallback(UART_HandleTypeDef *huart);

/* USER CODE BEGIN PFP */
// Private function prototypes

/* USER CODE END PFP */
```

```c
/* USER CODE BEGIN 0 */

/* USER CODE END 0 */

int main(void)
{

  /* USER CODE BEGIN 1 */

  /* USER CODE END 1 */

  // MCU Configuration

  // Reset all peripherals, initializes the flash and the Systick.
  HAL_Init();

  /* USER CODE BEGIN Init */

  /* USER CODE END Init */

  /* Configure the system clock */
  SystemClock_Config();

  /* USER CODE BEGIN SysInit */

  /* USER CODE END SysInit */

  /* Initialize all configured peripherals */
  MX_GPIO_Init();
  MX_USART2_UART_Init();

  /* USER CODE BEGIN 2 */
  HAL_UART_Receive_IT(&huart2, Rx_data,1);

  /* USER CODE END 2 */

  /* Infinite loop */
  /* USER CODE BEGIN WHILE */
  while (1)
  {
  /* USER CODE END WHILE */

  /* USER CODE BEGIN 3 */

      }
  /* USER CODE END 3 */

}

// System Clock Configuration
```

```
void SystemClock_Config(void)
{
  RCC_OscInitTypeDef RCC_OscInitStruct;
  RCC_ClkInitTypeDef RCC_ClkInitStruct;

  // Initialize the CPU, AHB and APB buss clocks
  RCC_OscInitStruct.OscillatorType = RCC_OSCILLATORTYPE_HSI;
  RCC_OscInitStruct.HSIState = RCC_HSI_ON;
  RCC_OscInitStruct.HSICalibrationValue = 16;
  RCC_OscInitStruct.PLL.PLLState = RCC_PLL_ON;
  RCC_OscInitStruct.PLL.PLLSource = RCC_PLLSOURCE_HSI;
  RCC_OscInitStruct.PLL.PLLMUL = RCC_PLL_MUL16;
  if (HAL_RCC_OscConfig(&RCC_OscInitStruct) != HAL_OK)
  {
    _Error_Handler(__FILE__, __LINE__);
  }

  // Initializes the CPU, AHB and APB buss clocks

  RCC_ClkInitStruct.ClockType = RCC_CLOCKTYPE_HCLK|RCC_CLOCKTYPE_SYSCLK
                             |RCC_CLOCKTYPE_PCLK1|RCC_CLOCKTYPE_PCLK2;
  RCC_ClkInitStruct.SYSCLKSource = RCC_SYSCLKSOURCE_PLLCLK;
  RCC_ClkInitStruct.AHBCLKDivider = RCC_SYSCLK_DIV1;
  RCC_ClkInitStruct.APB1CLKDivider = RCC_HCLK_DIV2;
  RCC_ClkInitStruct.APB2CLKDivider = RCC_HCLK_DIV1;

  if (HAL_RCC_ClockConfig(&RCC_ClkInitStruct, FLASH_LATENCY_2) !=
HAL_OK)
  {
    _Error_Handler(__FILE__, __LINE__);
  }

  // Configure the Systick interrupt time
  HAL_SYSTICK_Config(HAL_RCC_GetHCLKFreq()/1000);

  // Configure the Systick
  HAL_SYSTICK_CLKSourceConfig(SYSTICK_CLKSOURCE_HCLK);

  /* SysTick_IRQn interrupt configuration */
  HAL_NVIC_SetPriority(SysTick_IRQn, 0, 0);
}

/* USART2 init function */
static void MX_USART2_UART_Init(void)
{

  huart2.Instance = USART2;
  huart2.Init.BaudRate = 9600;
  huart2.Init.WordLength = UART_WORDLENGTH_8B;
  huart2.Init.StopBits = UART_STOPBITS_1;
  huart2.Init.Parity = UART_PARITY_NONE;
```

```
    huart2.Init.Mode = UART_MODE_TX_RX;
    huart2.Init.HwFlowCtl = UART_HWCONTROL_NONE;
    huart2.Init.OverSampling = UART_OVERSAMPLING_16;
    huart2.Init.OneBitSampling = UART_ONE_BIT_SAMPLE_DISABLE;
    huart2.AdvancedInit.AdvFeatureInit = UART_ADVFEATURE_NO_INIT;
    if (HAL_UART_Init(&huart2) != HAL_OK)
    {
        _Error_Handler(__FILE__, __LINE__);
    }

}
// This is the interrupt callback method
void HAL_UART_RxCpltCallback(UART_HandleTypeDef *huart) {
    uint8_t i;
    if (huart->Instance == USART2) // current UART
    {
        if (Rx_indx == 0)
        {
            for(i = 0; i < 100; i++)
              Rx_Buffer[i] = 0;  // clear Rx_Buffer prior to use
        // reset the user LED if previously set
                HAL_GPIO_WritePin(LD2_GPIO_Port, LD2_Pin, GPIO_PIN_RESET);
        }
        if (Rx_data[0] != 13)
        {
                Rx_Buffer[Rx_indx++] = Rx_data[0]; // add data to Rx_Buffer
        }
        else  // if received data = 13, (carriage return)
        {
                Rx_indx = 0;
                Transfer_cplt = 1; // transfer complete, data is ready
                HAL_UART_Transmit(&huart2, "\n\r", 2, 100);
                HAL_UART_Transmit(&huart2, Rx_Buffer, strlen(Rx_Buffer), 100);
                HAL_UART_Transmit(&huart2, "\n\r", 2, 100);
                if(!strcmp(Rx_Buffer,"LD2 on")) // LED trigger phrase
                {
                    HAL_GPIO_WritePin(LD2_GPIO_Port, LD2_Pin, GPIO_PIN_SET);
                }
        }

    HAL_UART_Receive_IT(&huart2, Rx_data, 1); // activate UART
receive interrupt
    HAL_UART_Transmit(&huart2, Rx_data, strlen(Rx_data), 100);
    }
}


/** Configure pins as
        * Analog
        * Input
        * Output
```

```
        * EVENT_OUT
        * EXTI
*/
static void MX_GPIO_Init(void)
{
  GPIO_InitTypeDef GPIO_InitStruct;

  /* GPIO Ports Clock Enable */
  __HAL_RCC_GPIOC_CLK_ENABLE();
  __HAL_RCC_GPIOF_CLK_ENABLE();
  __HAL_RCC_GPIOA_CLK_ENABLE();
  __HAL_RCC_GPIOB_CLK_ENABLE();

  /*Configure GPIO pin Output Level */
  HAL_GPIO_WritePin(LD2_GPIO_Port, LD2_Pin, GPIO_PIN_RESET);

  /*Configure GPIO pin : B1_Pin */
  GPIO_InitStruct.Pin = B1_Pin;
  GPIO_InitStruct.Mode = GPIO_MODE_IT_FALLING;
  GPIO_InitStruct.Pull = GPIO_NOPULL;
  HAL_GPIO_Init(B1_GPIO_Port, &GPIO_InitStruct);

  /*Configure GPIO pin : LD2_Pin */
  GPIO_InitStruct.Pin = LD2_Pin;
  GPIO_InitStruct.Mode = GPIO_MODE_OUTPUT_PP;
  GPIO_InitStruct.Pull = GPIO_NOPULL;
  GPIO_InitStruct.Speed = GPIO_SPEED_FREQ_LOW;
  HAL_GPIO_Init(LD2_GPIO_Port, &GPIO_InitStruct);

}

/* USER CODE BEGIN 4 */

/* USER CODE END 4 */

/**
  * @brief  This function is executed in case of error occurrence.
  * @param  None
  * @retval None
  */
void _Error_Handler(char * file, int line)
{
  /* USER CODE BEGIN Error_Handler_Debug */
  /* User can add his own implementation to report the HAL error
return state */
  while(1)
  {
  }
  /* USER CODE END Error_Handler_Debug */
}
```

```
#ifdef USE_FULL_ASSERT

/**
    * @brief Reports the name of the source file and the source line
number
    * where the assert_param error has occurred.
    * @param file: pointer to the source file name
    * @param line: assert_param error line source number
    * @retval None
    */
void assert_failed(uint8_t* file, uint32_t line)
{
  /* USER CODE BEGIN 6 */
  /* User can add his own implementation to report the file name and
line number,
      ex: printf("Wrong parameters value: file %s on line %d\r\n", file,
line) */
  /* USER CODE END 6 */

}

#endif

/**
  * @}
  */

/**
  * @}
*/

/**** (C) COPYRIGHT STMicroelectronics ****/
```

Test Run

The program was then built and downloaded into the project board. I then started the Realterm terminal program to display the communication stream from the board. Figure 8-9 shows the terminal output for the UART receive/transmit program.

I did note that the user LED did light when the phrase "LD2 on" was entered. The LED went off when the next character was entered as was expected.

This program will also be used as a template for other book projects that need both terminal input and output.

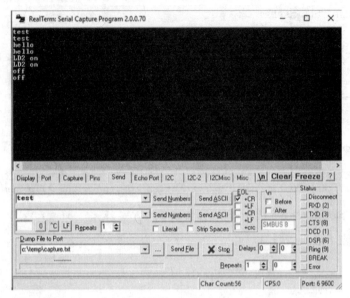

Figure 8-9 *Terminal output for the UART receive/transmit program.*

Summary

I began the chapter with an explanation of both the universal asynchronous receiver transmitter (UART) and the universal synchronous/asynchronous receiver/transmitter (USART) internal peripherals. This discussion also included how the data transfer protocol worked.

A discussion on how to configure a USART followed. The configuration procedure used two C structs to configure and initialize the USART. I also explained all the elements in both C structs.

I next showed how to install a terminal program, which is required to display alphanumeric data sent out by the USART. I selected the open-source terminal program named Realterm.

An explanation followed on how to enable the USART2 peripheral using the CubeMX application. The USART2 selection was configured in the CubeMX tree pane view. I also went through the procedure on how to configure the device using the C struct.

A demonstration program was next shown, which displayed a numeric variable that was continuously being updated.

The final demonstration displayed manually entered terminal data as well as lighting an LED when a trigger phrase was entered.

9

Analog-to-Digital
Conversion

This chapter covers a very important topic of analog-to-digital conversion. The devices that actually do the conversions are called analog-to-digital converters (ADCs). The STM line of MCU has a good variety of ADCs; however, they all operate and are programmed in a similar manner. There is a single ADC used in the project board STM32F302R8 MCU. However, it can accept analog inputs from any 1 of 15 separate external lines. There is a sixteenth ADC input, but it is permanently connected to an internal temperature sensor and is not available for any other use.

Beginners in the embedded development sometimes question the need for an ADC peripheral. The answer is that measurements in the real world are taken typically using analog sensors that provide a continuous range of voltages that are proportional to the environmental parameter that is being monitored. For instance, the Analog Devices TMP36 temperature sensor, shown in Figure 9-1, provides a range of voltages from approximately 0.1 to 2.0 V that is directly proportional to a range of temperatures roughly from −40°C to 150°C. This analog voltage must be converted to an equivalent digital number before being able to be processed by the MCU. That is the role of an ADC.

The following discussion provides a detailed explanation of how a STM ADC functions.

Figure 9-1 *Analog Devices TMP36 temperature sensor.*

ADC Functions

The technique used in STM ADCs is known as successive–approximation register (SAR). The architecture for a SAR ADC is fairly simple as may be seen in the block diagram, as shown in Figure 9-2.

An analog voltage is first inputted to a track/hold circuit. This circuit is also commonly called a sample/hold circuit, which takes an instantaneous sample of the varying analog voltage and uses that sample as the value to be converted into a digital number. The N-BIT REGISTER block, shown in Figure 9-2, is set at a mid-scale value. This value depends upon the number of bits created in the final conversion. For this discussion, I will use only 4 bits for the conversion, which will be adequate to explain the conversion process. In reality, the real STM ADC uses 12 bits. Getting back to the register will mean a mid-scale value of 1000 will be preset in the register. This value will therefore force the digital-to-analog converter (DAC) shown in the figure as N-BIT DAC to output a voltage equal to $V_{REF}/2$. V_{REF} is a reference voltage used by the ADC and is often the same as the supply or power rail voltage. That would be 3.3 V for most STM MCUs.

A comparison between the sampled analog voltage held by the TRACK/HOLD (V_{IN}) module and the DAC output (V_{DAC}) is then accomplished by a

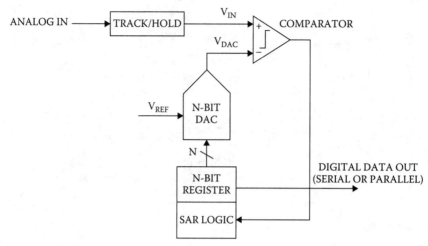

Figure 9-2 *SAR ADC block diagram.*

COMPARATOR module shown in the upper right-hand portion of Figure 9-2. The COMPARATOR will output a 1 if V_{IN} is greater than V_{DAC}, otherwise it will output a 0. If a 1 is output, the most significant bit in the N-BIT REGISTER will be retained, otherwise it will be reset to 0.

The SAR LOGIC module will then shift 1 bit to the right and repeat the comparison process. This sequence is repeated until the least significant bit (LSB) position is reached and processed. Once that point has been done, the N-BIT REGISTER will then contain the complete digital equivalent to the sampled analog input voltage. The SAR LOGIC module will then allow the fully converted digital value to be released either in a parallel or serial format.

Figure 9-3 shows a complete 4-bit conversion sequence.

The vertical axis represents the V_{DAC} voltage and the horizontal axis represents time. The result of the first conversion is to set the MSB to 0 because V_{IN} is less than $V_{REF}/2$. The SAR LOGIC module will then reset the V_{DAC} to $V_{REF}/4$ before the next comparison. In this case, V_{IN} is greater than $V_{REF}/4$ and the bit one position to the right of the MSB will be set to 1. Next, the SAR LOGIC module will split the difference between $V_{REF}/2$ and $V_{REF}/4$, which is $3V_{REF}/8$. This new comparison results in a 0 being output from the COMPARATOR module. In actuality, the SAR LOGIC module changes the VDAC voltage in some multiple of $V_{REF}/2^4$ or $V_{REF}/16$ for each comparison operation. This results in the converging step-like waveform, as shown in Figure 9-3. Some multiples of the step value are either

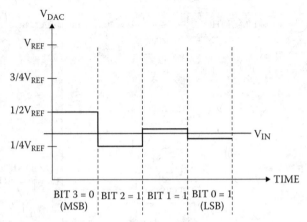

Figure 9-3 *Four-bit conversion sequence.*

added to or subtracted from V_{DAC} depending upon the result of the previous bit comparison. In this case the resulting digital value held by the N-BIT REGISTER is 0101, which is 5 in a decimal format. This all makes sense if you equate V_{REF} to 16 where each step in the V_{REF} range is 1. The V_{IN} line in the figure would then be approximately at the 5 level.

I believe you can see that the SAR process converges fairly rapidly to a final value. In this case, it took four comparisons, while a 12-bit ADC would take 12 comparisons. The good news is the SAR LOGIC module operations are all done in firmware with no MCU instructions required other than to start the conversion.

ADC Module with HAL

The HAL framework uses a similar C struct to declare ADC peripherals, as has been noted earlier when discussing GPIO pins and timer peripherals. The following C struct is defined for an ADC. Each struct member will separately be discussed after the struct definition.

```
typedef struct {
    ADC_TypeDef           *Instance;
    ADC_InitTypeDef       Init;
    __IO uint32_t         NbrOfCurrentConversionRank;
    DMA_HandleTypeDef     *DMA_Handle;
    HAL_LockTypeDef       Lock;
```

```
    __IO  uint32_t       State;
    __IO uint32_t        ErrorCode;
} ADC_HandleTypeDef;
```

The `ADC_HandleTypeDef` struct members are briefly defined in the following list:

`ADC_TypeDef *Instance;`—This is a pointer to the active ADC. For example, ADC1 would reference the first ADC.

`ADC_InitTypeDef Init;`—References the ADC_InitTypeDef C struct, which is used to configure the selected ADC. This struct will be discussed shortly.

`__IO uint32_t NbrOfCurrentConversionRank;`—References the rank or channel number in a conversion group. This will be shortly explained in detail.

`DMA_HandleTypeDef *DMA_Handle;`—This is a pointer to a DMA handler used for ADC to DMA conversions. This will be explained in the ADC/DMA section.

`HAL_LockTypeDef Lock;`—A locking object used for the ADC process.

`__IO uint32_t State;`—A state object used with the ADC communication process.

`__IO uint32_t ErrorCode;`—An object that can hold an error value that might arise.

An ADC declared using the `ADC_HandleTypeDef` struct must further be configured using the `ADC_InitTypeDef` C struct, which is defined as follows:

```
typedef struct {
uint32_t ClockPrescaler;
uint32_t Resolution;
uint32_t DataAlign;
uint32_t ScanConvMode;
uint32_t EOCSelection
uint32_t ContinuousConvMode;
uint32_t NbrOfConversion;
uint32_t DiscontinuousConvMode;
uint32_t NbrOfDiscConversion;
```

```
uint32_t ExternalTrigConv;
uint32_t ExternalTrigConvEdge;
uint32_t DMAContinuousRequests;
} ADC_InitTypeDef;
```

The ADC_InitTypeDef struct members are defined in the following list:

uint32_t ClockPrescaler;—This defines the ADC clock speed (ADC-CLK) used in the ADC. This clock rate indirectly sets the maximum sample rate possible with the ADC. The ADC prescaler has preset divider rates starting at one and progressing at multiples of 2, that is, 2, 4, 6, or 8. The ADC clock rate affects all MCU ADCs. This parameter can be any value of the following ADC_ClockPrescaler defines:

```
ADC_CLOCK_SYNC_PCLK_DIV1
ADC_CLOCK_SYNC_PCLK_DIV2
ADC_CLOCK_SYNC_PCLK_DIV4
ADC_CLOCK_SYNC_PCLK_DIV6
ADC_CLOCK_SYNC_PCLK_DIV8
```

uint32_t Resolution;—This value sets the ADC resolution. This parameter can be any value of the following ADC_Resolution defines:

```
ADC_RESOLUTION_12B - 12 bits
ADC_RESOLUTION_6B - 6 bits
```

There is a definite relationship between the resolution and maximum possible conversions that can be accomplished per second. This rule is *the greater the resolution, the lower the maximum conversion rate.*

uint32_t DataAlign;—This value establishes how the bits in the N-Bit register are aligned. This parameter can be any value of the following ADC_DATAALIGN defines:

```
ADC_DATAALIGN_LEFT—MSB to the left
ADC_DATAALIGN_RIGHT—MSB to the right
```

uint32_t ScanConvMode;—This value specifies the ADC mode. It is either single or continuous. This parameter can be either value of the following ADC_SCAN defines:

```
ADC_SCAN_DISABLE—Scan mode disabled
ADC_SCAN_ENABLE—Scan mode enabled
```

`uint32_t EOCSelection;`—This value sets the flag type indicating an end of a conversion (EOC). The flag type value also depends upon the ADC mode, which may be single or continuous. This parameter can be set to either of the following ADC_EOC defines:

```
ADC_EOC_SINGLE_CONV-Single conversions
ADC_EOC_SEQ_CONV-Continuous conversion
```

`uint32_t ContinuousConvMode`—Specifies whether the conversion is performed in single mode (one conversion) or continuous mode for regular group, after the selected trigger occurred (software start or external trigger). This parameter can be set to `ENABLE` or `DISABLE`.

`uint32_t NbrOfConversion;`—This value specifies the number of ranks or channels that will be converted within the regular group sequencer. To use regular group sequencer and convert several ranks, parameter 'ScanConvMode' must be enabled. This parameter must be a number between Min_Data = 1 and Max_Data = 16.

`uint32_t DiscontinuousConvMode;`—This value specifies whether the conversion sequence of regular group is performed in Complete sequence/ Discontinuous sequence (main sequence subdivided in successive parts). Discontinuous mode is used only if sequencer is enabled (parameter 'ScanConvMode'). If sequencer is disabled, this parameter is discarded. Discontinuous mode can be enabled only if continuous mode is disabled. If continuous mode is enabled, this parameter setting is discarded. This parameter can be set to `ENABLE` or `DISABLE`.

`uint32_t NbrOfDiscConversion;`—This value specifies the number of discontinuous conversions in which the main sequence of regular group (parameter NbrOfConversion) will be subdivided. If parameter 'DiscontinuousConv-Mode' is disabled, this parameter is discarded. This parameter must be a number between Min_Data = 1 and Max_Data = 8.

`uint32_t ExternalTrigConv;`—This value selects the external event used to trigger the conversion start of regular group. If set to ADC_SOFTWARE_ START, external triggers are disabled. If set to external trigger source, triggering is on event-rising edge by default. This parameter can be a value in the format of ADC_External_trigger_Source_Regular and is one of the following defines:

```
ADC_EXTERNALTRIGCONV_T1_CC1
ADC_EXTERNALTRIGCONV_T1_CC2
ADC_EXTERNALTRIGCONV_T1_CC3
ADC_EXTERNALTRIGCONV_T2_CC2
ADC_EXTERNALTRIGCONV_T5_TRGO
ADC_EXTERNALTRIGCONV_T4_CC4
ADC_EXTERNALTRIGCONV_T3_CC4
ADC_EXTERNALTRIGCONV_T8_TRGO
ADC_EXTERNALTRIGCONV_T8_TRGO2
ADC_EXTERNALTRIGCONV_T1_TRGO
ADC_EXTERNALTRIGCONV_T1_TRGO2
ADC_EXTERNALTRIGCONV_T2_TRGO
ADC_EXTERNALTRIGCONV_T4_TRGO
ADC_EXTERNALTRIGCONV_T6_TRGO
ADC_EXTERNALTRIGCONV_EXT_IT11
ADC_SOFTWARE_START
```

`uint32_t ExternalTrigConvEdge;` —This value selects the external trigger edge of regular group. If trigger is set to ADC_SOFTWARE_START, this parameter is discarded. This parameter can be a value in the format of ADC_External_trigger_edge_Regular and is one of the following defines:

```
ADC_EXTERNALTRIGCONVEDGE_NONE
ADC_EXTERNALTRIGCONVEDGE_RISING
ADC_EXTERNALTRIGCONVEDGE_FALLING
ADC_EXTERNALTRIGCONVEDGE_RISINGFALLING
```

`uint32_t DMAContinuousRequests;` —This value specifies whether the DMA requests are performed in one shot mode (DMA transfer stop when number of conversions is reached) or in continuous mode (DMA transfer unlimited, whatever number of conversions). Note: In continuous mode, DMA must be configured in circular mode. Otherwise, an overrun will be triggered when DMA buffer maximum pointer is reached. Note: This parameter must be modified when no conversion is ongoing on both regular and injected groups (ADC disabled, or ADC enabled without continuous mode or external trigger that could launch a conversion). This parameter can be set to ENABLE or DISABLE.

Don't be too discouraged by the extent of the above listings. The actual declaration and configuration of a STM ADC peripheral is straightforward and reasonably easy to understand as you will shortly realize by examining a demonstration.

However, I still need to cover some remaining ADC topics before proceeding to a real demonstration.

ADC Conversion Modes

There are few common ways to use one or more input channels with an ADC. These ways are known as conversion modes and I will cover the most popular conversion modes next.

Single Channel/Single Conversion

This is the absolute simplest mode. Figure 9-4 shows a block diagram for this mode.

In this mode, the ADC takes or more accurately samples the analog voltage present on a selected input line or channel and converts that voltage to a digital number. The number is then read from the ADC and used in any application that requires it.

Multi-channel Scan/Single Conversion

This mode is a little more complex than the previous one. In this mode, multiple input analog lines are sampled and the sampled voltage levels are then converted into digital numbers. The block diagram for this mode is shown in Figure 9-5.

One of the exceptionally nice features for this mode is that each channel is independent and may be configured to have different sample rates, trigger sources, and other custom items. This feature is implemented by using a ranks object. By using ranks, the MCU does not have to stop and reconfigure each channel for its

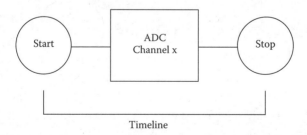

Figure 9-4   *Single channel/single conversion ADC mode block diagram.*

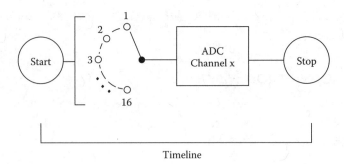

Figure 9-5 *Multi-channel scan/single conversion ADC mode block diagram.*

custom configuration. This mode only does one scan and the result numbers are often directly stored in a memory using a direct memory access (DMA).

Single Channel/Continuous Conversion

In this mode, a single channel is continuously being sampled and the resulting digital numbers are continuously being stored in memory. This mode almost always uses DMA to alleviate excessive load on the MCU. This mode is ordinarily use for real-time monitoring of sensors or other environmental devices.

Multi-Channel Scan/Continuous Conversion

In this mode, multiple channels are being continuously sampled and the resulting digital numbers are stored in memory. This mode demands that DMA be used to alleviate an intolerable computation load on the MCU. This mode is ordinarily used for comprehensive real-time monitoring of sensors or other environmental devices.

Channels, Groups, and Ranks

The number of channels that can be converted varies with the particular STM MCU being used. That number is set at 16 for the STM32F302R8 MCU used in the Nucleo project board. Sequences of channel conversions may be set up within independent collections known as groups. Channels may be arranged in any order within a specific group.

Input channels are hardware based and thus fixed and linked to specific MCU pins, that is, IN0 is the first channel, IN1 the second, and so on. They can,

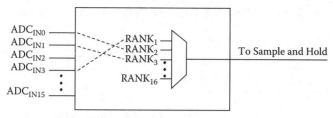

Figure 9-6 *ADC diagram showing channels and ranks.*

however, be logically reordered to form a custom sampling sequence within a group. The reordering of channels is done by assigning the input channel to an index ranging from 1 to 16. This index is called rank in the HAL.

Figure 9-6 shows this concept. Although the IN3 channel in the figure is fixed, for example, it is connected to PA3 pin in an STM32F302R8 MCU, it can be assigned to the rank 1 to make it the first channel to be sampled.

The following C struct is defined for an `ADC_ChannelConfTypeDef`. Each struct member will separately be discussed after the struct definition.

```
typedef struct {
uint32_t Channel;
uint32_t Rank;
uint32_t SamplingTime;
uint32_t Offset;
} ADC_ChannelConfTypeDef;
```

The `ADC_ChannelConfTypeDef` struct members are defined in the following list:

`uint32_t Channel;`—This value specifies the channel to configure into ADC regular group. This parameter can be any value of the following ADC_CHAN-NEL defines:

```
ADC_CHANNEL_0
to
ADC_CHANNEL_15
ADC_CHANNEL_VREFINT
ADC_CHANNEL_VBAT
```

`uint32_t Rank;`—This value specifies the rank in the regular group sequencer. This parameter must be a number between Min_Data = 1 and Max_Data = 16:

This parameter can be any value of the following ADC_REGULAR_RANK defines:

```
ADC_REGULAR_RANK_1
to
ADC_REGULAR_RANK16
```

`uint32_t SamplingTime;`—The sampling time value to be set for the selected channel in units of ADC clock cycles. The total conversion time is the addition of sampling time and processing time (12 ADC clock cycles at ADC resolution 12 bits, 11 cycles at 10 bits, 9 cycles at 8 bits, 7 cycles at 6 bits). This parameter can be any value of the following ADC_SAMPLETIME defines:

```
ADC_SAMPLETIME_3CYCLES
ADC_SAMPLETIME_15CYCLES
ADC_SAMPLETIME_28CYCLES
ADC_SAMPLETIME_56CYCLES
ADC_SAMPLETIME_84CYCLES
ADC_SAMPLETIME_112CYCLES
ADC_SAMPLETIME_144CYCLES
ADC_SAMPLETIME_480CYCLES
```

CAUTION: This parameter updates the parameter property of the channel, which can be used into regular and/or injected groups. If this same channel has been previously configured in the other group (regular/injected), it will be updated to last setting. In case of usage of internal measurement channels (VrefInt/Vbat/TempSensor), sampling time constraints must be respected (sampling time can be adjusted in function of ADC clock frequency and sampling time setting).

`uint32_t Offset;`—Reserved for future use, can be set to 0.

ADC Demonstration

The ADC, like many STM peripherals, can be controlled using one of three methods:

- Polling
- Interrupt
- DMA

This first demonstration will use polling as the control approach because it is the simplest and will allow the focus to be mainly on the ADC process.

The initial step is to create a reference to a `ADC_HandleTypeDef` struct and then assign the appropriate values to struct members that will define the ADC instance to be implemented. This new instance will next be configured using the `ADC_InitTypeDef` struct, which requires a reference to the now defined `ADC_HandleTypeDef` struct. It will time to initialize and start the ADC once all the `ADC_InitTypeDef` struct members have been defined. The initialize and start commands are accomplished using these calls:

```
HAL_ADC_Init()
HAL_ADC_Start()
```

The ADC will then either do one conversion if the single mode was selected or continue with a stream of conversions if the continuous mode was selected. In the case of the single mode, the `HAL_ADC_Stop()` method must first be called to stop the ADC thread and then the `HAL_ADC_Start()` method must be called again to do another conversion.

Any application desiring access to the converted data must use the following method when using polling to determine when an ADC conversion has finished and the data is available:

```
HAL_StatusTypeDef HAL_ADC_PollForConversion(ADC_Handle-
TypeDef* hadc, uint32_t Timeout);
```

The argument methods are a pointer to the `ADC_HandleTypeDef` struct and an integer named Timeout representing the maximum number of milliseconds to wait for data to be available from the ADC register. It is also possible to use the define HAL_MAX_DELAY for the Timeout value, which will cause the system to wait indefinitely for data. I would not recommend that approach because it has the potential to "hang" the system without any apparent symptom.

You must use the following method to actually transfer the data from the ADC register to any user application:

```
uint32_t HAL_ADC_GetValue(ADC_HandleTypeDef* hadc);
```

It is now appropriate to discuss the Analog Devices TMP36 temperature sensor, which I introduced at the chapter's start. This sensor will be the analog voltage source used for the ADC demonstration.

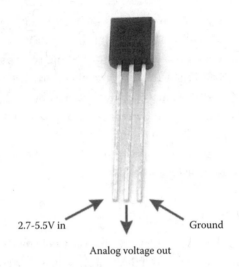

2.7-5.5V in   Ground

Analog voltage out

Figure 9-7   *TMP36 lead connections.*

Analog Devices TMP36 Temperature Sensor

An image of this sensor has been shown in Figure 9-1. It is housed in a regular TO-92 plastic case with three leads as shown in the figure. It does, however, contain quite a bit of internal components that allow it to output a DC voltage that is proportional to the ambient temperature surrounding its case. This sensor is also very inexpensive and is readily available from a number of online sources. The sensor specifications are listed below:

- Size: TO-92 package (about 0.2″ × 0.2″ × 0.2″) with three leads

- Temperature range: −40°C to 150°C/−40°F to 302°F

- Output range: 0.1V (−40°C) to 2.0 V (150°C) (decreasing accuracy after 125°C)

- Power supply: 2.7 V to 5.5 V only, with a typical 0.05-mA current draw

Figure 9-7 shows to what each lead should be connected.

The leads should be connected to the Arduino protoboard, as detailed in Table 9-1.

The temperature is very easy to calculate using the following equation:

$$°C = [(\text{Vout in mV}) - 500] / 10$$

TMP36 lead	Arduino protoboard (refer to Figure 5-11)
Power	CN6, pin 5 (+5 V)
Ground	CN6, pin 6 (GND)
Output	CN8, pin 1 (PA0)

Table 9-1   *TMP36 Lead Connections*

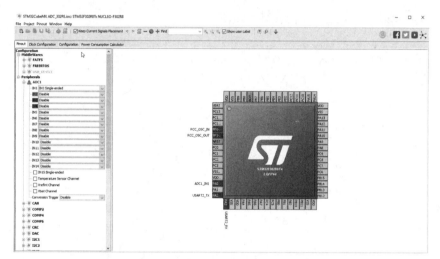

Figure 9-8   *Enabling the IN1 channel in ADC1.*

The equivalent temperature in °F is also easily calculated using this conversion equation:

$$°F = (9 * °C/5) + 32$$

It is now time to discuss the main.c file as all the precursor items have been introduced and explained.

Enabling ADC1

I will now need to enable the ADC1 alternative function in order to have the analog-to-digital conversion function. You configure ADC1 when creating a new project and the Pinout window is displayed. Figure 9-8 shows how the ADC1 peripheral is enabled by selecting IN1 Single-ended from the ADC1 IN1 drop-down menu.

The CubeMX application will automatically insert an initialization method static void MX_ADC1_Init(void) in the main.c file when the ADC1 functionality is enabled.

The appropriate ADC parameters are automatically set for you based upon the parameters you set in the ADC1 IP tree view, as shown in the following code snippet:

```
static void MX_ADC1_Init(void)
{
  ADC_ChannelConfTypeDef sConfig;
  // Common config
  hadc1.Instance = ADC1;
  hadc1.Init.ClockPrescaler = ADC_CLOCK_ASYNC_DIV1;
  hadc1.Init.Resolution = ADC_RESOLUTION_12B;
  hadc1.Init.ScanConvMode = ADC_SCAN_DISABLE;
  hadc1.Init.ContinuousConvMode = DISABLE;
  hadc1.Init.DiscontinuousConvMode = ENABLE;
  hadc1.Init.NbrOfDiscConversion = 1;
  hadc1.Init.ExternalTrigConvEdge = ADC_EXTERNALTRIGCONVEDGE_NONE;
  hadc1.Init.ExternalTrigConv = ADC_SOFTWARE_START;
  hadc1.Init.DataAlign = ADC_DATAALIGN_RIGHT;
  hadc1.Init.NbrOfConversion = 1;
  hadc1.Init.DMAContinuousRequests = DISABLE;
  hadc1.Init.EOCSelection = ADC_EOC_SINGLE_CONV;
  hadc1.Init.LowPowerAutoWait = DISABLE;
  hadc1.Init.Overrun = ADC_OVR_DATA_OVERWRITTEN;
  if (HAL_ADC_Init(&hadc1) != HAL_OK)
  {
    _Error_Handler(__FILE__, __LINE__);
  }
  // Configure Regular Channel
  sConfig.Channel = ADC_CHANNEL_1;
  sConfig.Rank = 1;
  sConfig.SingleDiff = ADC_SINGLE_ENDED;
  sConfig.SamplingTime = ADC_SAMPLETIME_1CYCLE_5;
  sConfig.OffsetNumber = ADC_OFFSET_NONE;
  sConfig.Offset = 0;
  if (HAL_ADC_ConfigChannel(&hadc1, &sConfig) != HAL_OK)
  {
    _Error_Handler(__FILE__, __LINE__);
  }
}
```

Connecting ADC1 to the TMP36 Sensor

Figure 9-9 shows how I connected the TMP36 temperature sensor to the ADC1 internal peripheral using a solderless breadboard, jumper leads, and the Arduino protoboard.

Figure 9-9 *Interconnections between the TMP36 sensor and the Arduino protoboard.*

The TMP36 lead connections have been previously detailed in Table 9-1. You will need three jumper leads to connect the sensor to the Arduino protoboard.

At this point, all the prerequisites have been introduced that are needed to demonstrate an actual ADC program.

ADC Demonstration Software

The following program is based on the USART program shown in the previous chapter. This step was taken because the processed ADC output is required to be displayed in a terminal window. The main.c file now has a new method that configures and initializes the ADC1 internal peripheral.

There is also new functional code that runs within the forever loop. This code is shown in the following code snippet with detailed explanations following the listing:

```
while (1)
{
/* USER CODE END WHILE */
/* USER CODE BEGIN 3 */
        HAL_ADC_Start(&hadc1);
        HAL_ADC_PollForConversion(&hadc1, 100);
        raw = (double) HAL_ADC_GetValue(&hadc1);
        raw = raw * 0.452
        value = (raw - 500.0)/10;
        value = (9*value)/5 + 32;
        sprintf(msg, "%f\r\n", value);
        HAL_UART_Transmit(&huart2, (uint8_t*) msg, 20, 100);
        HAL_Delay(1000);

}
/* USER CODE END 3 */
}
```

All the statements within the code snippet are discussed below:

- HAL_ADC_Start(&hadc1);—Starts the ADC1 internal peripheral.

- HAL_ADC_PollForConversion(&hadc1, 100);—Waits for an end-of-conversion (EOC) signal before proceeding to the next instruction. The 100 value is the maximum time in milliseconds to wait for the EOC signal. This is a blocking type instruction.

- raw = (double) HAL_ADC_GetValue(&hadc1);—Transfer the value from the adc output register to the raw variable. This is a 12-bit value with a maximum numerical value of 4095, which is roughly equivalent to 5 V. The value is also cast to a double format.

- raw = raw * 0.452;—Conversion from a raw count to an absolute mV output. The scaling factor 0.452 was determined using a precision voltage measurement.

- value = (raw - 500.0)/10;—The raw mV value is converted to a °C temperature using the TMP36 equation.

- value = (9*value)/5 + 32;—Classic equation to convert from °C to °F.

- `sprintf(msg, "%f\r\n", value);`—Causes the double variable `value` to be converted to a string with a carriage return and line feed concatenated to the value.

- `HAL_UART_Transmit(&huart2, (uint8_t*) msg, 20, 100);`—Performs the actual transmission of the data from the MCU to the terminal program

- `HAL_Delay(1000);`—Inserts a one second delay between samples.

The complete main.c program is listed below:

```
// STM disclaimer goes here
// Includes
#include "main.h"
#include "stm32f3xx_hal.h"
/* USER CODE BEGIN Includes */
/* USER CODE END Includes */
// Private variables
ADC_HandleTypeDef hadc1;
UART_HandleTypeDef huart2;
/* USER CODE BEGIN PV */
// Private variables
double value = 1.0;
double raw =0.0;
char msg[20] = "";
/* USER CODE END PV */
// Private function prototypes
void SystemClock_Config(void);
static void MX_GPIO_Init(void);
static void MX_USART2_UART_Init(void);
static void MX_ADC1_Init(void);
/* USER CODE BEGIN PFP */
// Private function prototypes
/* USER CODE END PFP */
/* USER CODE BEGIN 0 */
/* USER CODE END 0 */
int main(void)
{
  /* USER CODE BEGIN 1 */
  /* USER CODE END 1 */
  // MCU Configuration
  // Reset all peripherals, initializes the flash and Systick.
  HAL_Init();
  /* USER CODE BEGIN Init */
  /* USER CODE END Init */
  /* Configure the system clock */
  SystemClock_Config();
  /* USER CODE BEGIN SysInit */
  /* USER CODE END SysInit */
```

```
/* Initialize all configured peripherals */
MX_GPIO_Init();
MX_USART2_UART_Init();
MX_ADC1_Init();
/* USER CODE BEGIN 2 */
/* USER CODE END 2 */
/* Infinite loop */
/* USER CODE BEGIN WHILE */
while (1)
{
/* USER CODE END WHILE */
/* USER CODE BEGIN 3 */
        HAL_ADC_Start(&hadc1);
        HAL_ADC_PollForConversion(&hadc1, 100);
        raw = (double) HAL_ADC_GetValue(&hadc1);
        raw = raw 8 0.452;
        value = (raw - 500.0)/10;
        value = (9*value)/5 + 32;
        sprintf(msg, "%f\r\n", value);
        HAL_UART_Transmit(&huart2, (uint8_t*) msg, 20, 100);
        HAL_Delay(1000);
}
/* USER CODE END 3 */
}
// System Clock Configuration
void SystemClock_Config(void)
{
  RCC_OscInitTypeDef RCC_OscInitStruct;
  RCC_ClkInitTypeDef RCC_ClkInitStruct;
  RCC_PeriphCLKInitTypeDef PeriphClkInit;
  // Initialize the CPU, AHB and APB buss clocks
  RCC_OscInitStruct.OscillatorType = RCC_OSCILLATORTYPE_HSI;
  RCC_OscInitStruct.HSIState = RCC_HSI_ON;
  RCC_OscInitStruct.HSICalibrationValue = 16;
  RCC_OscInitStruct.PLL.PLLState = RCC_PLL_ON;
  RCC_OscInitStruct.PLL.PLLSource = RCC_PLLSOURCE_HSI;
  RCC_OscInitStruct.PLL.PLLMUL = RCC_PLL_MUL16;
  if (HAL_RCC_OscConfig(&RCC_OscInitStruct) != HAL_OK)
  {
    _Error_Handler(__FILE__, __LINE__);
  }
  // Initialize the CPU, AHB and APB buss clocks
  RCC_ClkInitStruct.ClockType =
  RCC_CLOCKTYPE_HCLK|RCC_CLOCKTYPE_SYSCLK
  |RCC_CLOCKTYPE_PCLK1|RCC_CLOCKTYPE_PCLK2;
  RCC_ClkInitStruct.SYSCLKSource = RCC_SYSCLKSOURCE_PLLCLK;
  RCC_ClkInitStruct.AHBCLKDivider = RCC_SYSCLK_DIV1;
  RCC_ClkInitStruct.APB1CLKDivider = RCC_HCLK_DIV2;
  RCC_ClkInitStruct.APB2CLKDivider = RCC_HCLK_DIV1;
  if (HAL_RCC_ClockConfig(&RCC_ClkInitStruct, FLASH_LATENCY_2) !=
HAL_OK)
  {
```

```
    _Error_Handler(__FILE__, __LINE__);
  }
  PeriphClkInit.PeriphClockSelection = RCC_PERIPHCLK_ADC1;
  PeriphClkInit.Adc1ClockSelection = RCC_ADC1PLLCLK_DIV1;
  if (HAL_RCCEx_PeriphCLKConfig(&PeriphClkInit) != HAL_OK)
  {
    _Error_Handler(__FILE__, __LINE__);
  }

  // Configure the Systick interrupt time
  HAL_SYSTICK_Config(HAL_RCC_GetHCLKFreq()/1000);
  // Configure the Systick
  HAL_SYSTICK_CLKSourceConfig(SYSTICK_CLKSOURCE_HCLK);
  /* SysTick_IRQn interrupt configuration */
  HAL_NVIC_SetPriority(SysTick_IRQn, 0, 0);
}
/* ADC1 init function */
static void MX_ADC1_Init(void)
{
  ADC_ChannelConfTypeDef sConfig;
  // Common config
  hadc1.Instance = ADC1;
  hadc1.Init.ClockPrescaler = ADC_CLOCK_ASYNC_DIV1;
  hadc1.Init.Resolution = ADC_RESOLUTION_12B;
  hadc1.Init.ScanConvMode = ADC_SCAN_DISABLE;
  hadc1.Init.ContinuousConvMode = DISABLE;
  hadc1.Init.DiscontinuousConvMode = ENABLE;
  hadc1.Init.NbrOfDiscConversion = 1;
  hadc1.Init.ExternalTrigConvEdge = ADC_EXTERNALTRIGCONVEDGE_NONE;
  hadc1.Init.ExternalTrigConv = ADC_SOFTWARE_START;
  hadc1.Init.DataAlign = ADC_DATAALIGN_RIGHT;
  hadc1.Init.NbrOfConversion = 1;
  hadc1.Init.DMAContinuousRequests = DISABLE;
  hadc1.Init.EOCSelection = ADC_EOC_SINGLE_CONV;
  hadc1.Init.LowPowerAutoWait = DISABLE;
  hadc1.Init.Overrun = ADC_OVR_DATA_OVERWRITTEN;
  if (HAL_ADC_Init(&hadc1) != HAL_OK)
  {
    _Error_Handler(__FILE__, __LINE__);
  }
  // Configure Regular Channel
  sConfig.Channel = ADC_CHANNEL_1;
  sConfig.Rank = 1;
  sConfig.SingleDiff = ADC_SINGLE_ENDED;
  sConfig.SamplingTime = ADC_SAMPLETIME_1CYCLE_5;
  sConfig.OffsetNumber = ADC_OFFSET_NONE;
  sConfig.Offset = 0;
  if (HAL_ADC_ConfigChannel(&hadc1, &sConfig) != HAL_OK)
  {
    _Error_Handler(__FILE__, __LINE__);
  }

}
```

```
/* USART2 init function */
static void MX_USART2_UART_Init(void)
{
  huart2.Instance = USART2;
  huart2.Init.BaudRate = 9600;
  huart2.Init.WordLength = UART_WORDLENGTH_8B;
  huart2.Init.StopBits = UART_STOPBITS_1;
  huart2.Init.Parity = UART_PARITY_NONE;
  huart2.Init.Mode = UART_MODE_TX_RX;
  huart2.Init.HwFlowCtl = UART_HWCONTROL_NONE;
  huart2.Init.OverSampling = UART_OVERSAMPLING_16;
  huart2.Init.OneBitSampling = UART_ONE_BIT_SAMPLE_DISABLE;
  huart2.AdvancedInit.AdvFeatureInit = UART_ADVFEATURE_NO_INIT;
  if (HAL_UART_Init(&huart2) != HAL_OK)
  {
    _Error_Handler(__FILE__, __LINE__);
  }
}
/** Configure pins as
        * Analog
        * Input
        * Output
        * EVENT_OUT
        * EXTI
*/
static void MX_GPIO_Init(void)
{
  /* GPIO Ports Clock Enable */
  __HAL_RCC_GPIOF_CLK_ENABLE();
  __HAL_RCC_GPIOA_CLK_ENABLE();
}
/* USER CODE BEGIN 4 */
/* USER CODE END 4 */
/**
  * @brief  This function is executed in case of error occurrence.
  * @param  None
  * @retval None
  */
void _Error_Handler(char * file, int line)
{
  /* USER CODE BEGIN Error_Handler_Debug */
  /* User can add his own implementation to report the HAL error re-
turn state */
  while(1)
  {
  }
  /* USER CODE END Error_Handler_Debug */
}
#ifdef USE_FULL_ASSERT
/**
    * @brief Reports the name of the source file and the source line
number
```

```
 * where the assert_param error has occurred.
 * @param file: pointer to the source file name
 * @param line: assert_param error line source number
 * @retval None
 */
void assert_failed(uint8_t* file, uint32_t line)
{
  /* USER CODE BEGIN 6 */
  /* User can add his own implementation to report the file name and
line number,
    ex: printf("Wrong parameters value: file %s on line %d\r\n", file,
line) */
  /* USER CODE END 6 */
}
#endif
/**
  * @}
  */
/**
  * @}
  */
/**** (C) COPYRIGHT STMicroelectronics ****/
```

Test Run

The program was then built and downloaded into the project board. I then started the Realterm terminal program to display the communication stream from the board. Figure 9-10 shows the terminal output.

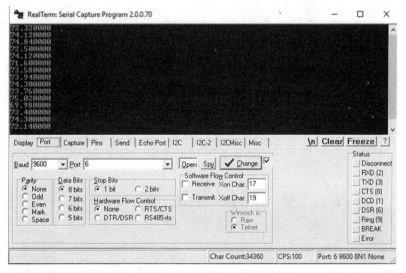

Figure 9-10   *Terminal output.*

I tried cooling and heating the sensor and saw the sensor reacting according to the ambient temperature.

Summary

Analog-to digital conversion (ADC) was the chapter's topic. There was an explanation of how the 16-channel, 12-bit ADC works within the MCU. It uses a successive approximation technique to acquire precise numerical samples of a continuously varying analog input.

The discussion next turned to how the HAL framework supports the ADC function. I went through the two C structs that both configure and initialize the ADC peripheral. The discussion also included the various conversion modes, channels, groups, and ranks.

A full-scale ADC demonstration closed out the chapter. A TMP36 temperature sensor was used as an analog voltage source to the ADC. The program used the USART2 peripheral to stream out continuous temperature readings to a terminal window.

10

Pulse Width Modulation (PWM)

This chapter concerns pulse width modulation (PWM), which is a digital signal easily generated by a MCU for a variety of purposes. These purposes commonly include controlling the following:

- A standard servo motor's position

- The rotational speed of a continuous rotation servo

- The luminescence of a light emitter

- The rotation speed of a standard electric motor with external driver circuitry

- A photovoltaic solar battery charger

- The maximum power-point track

- An output waveform such as sine, triangle, or square

- An acoustic sound generator

This chapter's projects will demonstrate how to control the luminescence (light intensity) of a normal LED, the color and luminescence of a tricolor LED, and the positioning of a standard servo motor.

It is wise to first discuss the nature and properties of a PWM signal before proceeding to show how to generate that using the STM project board. Figure 10-1 is a typical PWM signal with various properties annotated in the figure.

The first thing to note is that the signal is repetitive with a frequency around 50 Hz or an equivalent 20 ms period. This frequency is very common in servo

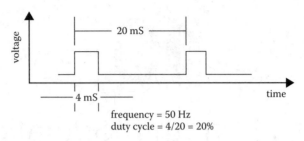

Figure 10-1 *Typical PWM signal.*

control and it will be what I will use for the chapter projects. However, there is nothing limiting about this frequency and you can use any value up to the limit of what the MCU can support.

The second important feature is the duty cycle, which is a percentage of the amount of time the signal is in a high state to the total period, which is the sum of the high and low times. Duty cycle is often considered the primary property for a PWM signal, for it solely determines the servo position and LED luminescence.

It turns out that each type of device being controlled by PWM acts differently with the signal. I will describe each specific interaction when I go through the device demonstration.

General-Purpose Timer PWM Signal Generation

PWM signals are generated only by using either an advanced or a general purpose (GP) timer. This section describes the process on how this is done.

The signal generation is relatively simple and involves using a timer in an upcounting mode. The top portion of Figure 10-2 shows a timer counting up from 0 to a maximum 16-bit count equal to 65535.

Each GP timer has a series of registers known as compare capture registers (CCRx), which store a number. When the upcounting reaches that number, an event is generated that depends on the timer mode. If the timer is set for a direct output as is the case for this PWM mode, then the event will be to turn on the channel associated with the CCRx register. In Figure 10-2, channel 1 is switched high when the count reaches 52428. The channel remains high until the timer counter reaches the maximum count (MAX) and overflows or resets as shown in the top waveform. The net result of this action is that a repetitive pulse is emitted

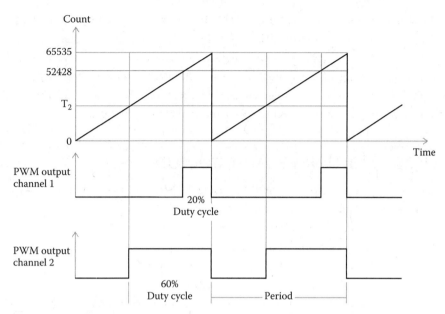

Figure 10-2 *PWM signal generation.*

from the channel 1 line with a 20% duty cycle. The following equation relates the timer count with the duty cycle (DC):

$$DC = \frac{MAX - CCR}{MAX}$$

This equation may be rearranged to find the CCRx value given a desired duty cycle:

$$CCR = MAX * (1 - DC)$$

In Figure 10-2, I have also shown an additional channel output with a desired 60% duty cycle but no specific count shown for the CCRx register. I will use the second form of the equation to solve for T_2, which is the count value to be stored in register CCR2:

$$CCR2 = 65535 * (1 - 0.6) = 26214$$

Therefore, storing a value of 26214 in CCR2 will cause a 60% duty cycle pulse train to be generated from TIM2's channel 2 output. Please note that the numbers I used for this calculation are relative in the sense that the actual CCRx numbers stored depend upon the real timer clock rate. You will shortly see that the numbers are much smaller, but the principle to figure them out is the same.

The TIM2 GP timer I used for PWM generation has a maximum of four output channels, meaning that it is possible to simultaneously generate four synchronized PWM signals using a single GP timer. One of the following demonstrations uses three channels, which is ample to demonstrate this capability.

Timer Hardware Architecture

Figure 10-3 is a portion of the GP timer block diagram highlighting the multi-channel outputs and associated CCRx registers.

The architecture is very straightforward, showing that each CCRx register has an input from the common timer counter (CNT) register as well as its own prescaled clock-rate input. This arrangement enables a very flexible PWM pulse train output to be implemented.

It is now time to describe how the HAL software enables PWM signal generation.

PWM Signals with HAL

The HAL framework uses a C struct named `TIM_MasterConfigTypeDef` to configure the timer peripheral generating a PWM signal. Each member will separately be discussed after the struct definition.

```
typedef struct {
uint32_t MasterOutputTrigger; /* Trigger output (TRGO) selection */
uint32_t MasterSlaveMode; /* Master/slave mode selection */
} TIM_MasterConfigTypeDef;
```

The `TIM_MasterConfigTypeDef` struct members are briefly defined in the following list:

`uint32_t MasterOutputTrigger;`—This value specifies the behavior of the trigger output (TRGO). This parameter can be any value of the following defines:

`uint32_t MasterSlaveMode;`—This is used to enable/disable the master/slave mode of a timer. This parameter can be one of the two values of the following defines:

```
TIM_MASTERSLAVEMODE_ENABLE
TIM_MASTERSLAVEMODE_DISABLE
```

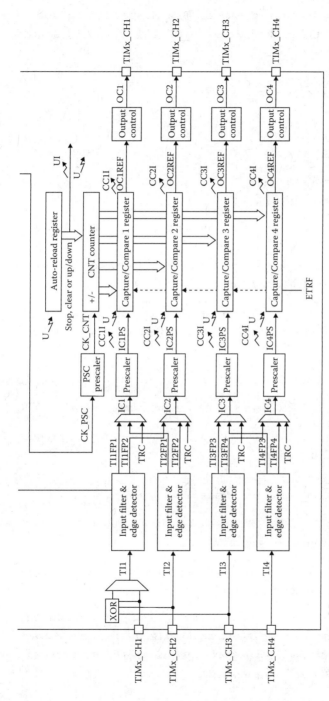

Figure 10-3 *GP timer block diagram with four output channels.*

A PWM timer configured using the `TIM_MasterConfigTypeDef` struct must further be initialized using the `TIM_OC_InitTypeDef` C struct, which is defined as follows:

```
typedef struct {
uint32_t OCMode;
uint32_t Pulse;
uint32_t OCPolarity;
uint32_t OCNPolarity;
uint32_t OCFastMode;
uint32_t OCIdleState;
uint32_t OCNIdleState;
} TIM_OC_InitTypeDef;
```

The `TIM_OC_InitTypeDef` struct members are briefly defined in the following list:

`uint32_t OCMode;`—Specifies the output compare mode. This parameter can be any value of the following defines:

`TIM_OCMODE_TIMING`—The comparison between the output compare register (CCRx) and the timer counter (CNT) has no effect on the output, which is known as a frozen mode.

`TIM_OCMODE_ACTIVE`—Set the channel output to active level on CCRx match.

`TIM_OCMODE_INACTIVE`—Set channel output to inactive level on CCRx match.

`TIM_OCMODE_TOGGLE`—Toggle channel output when the timer counter (CNT) matches the CCRx.

```
TIM_OCMODE_PWM1—PWM Mode 1
TIM_OCMODE_PWM2—PWM Mode 2
```

`TIM_OCMODE_FORCED_ACTIVE`—Force channel output high independently from the timer counter (CNT) value.

`TIM_OCMODE_FORCED_INACTIVE`—Force channel output low independently from the timer counter (CNT) value.

NOTE *PWM Mode 1: When the timer is upcounting, the channel will be active for as long as the Period is less than the Pulse, otherwise it is inactive. In a downcounting mode, the channel is inactive for as long as the Period is greater than the Pulse, otherwise it is active.*

PWM Mode 2: When the timer is upcounting, the channel will be inactive as long as the Period is less than the Pulse, otherwise it is active. In downcounting mode, the channel is active for as long as the Period is greater than the Pulse, otherwise it is inactive.

`uint32_t Pulse;` —This value be stored inside the CCRx register and it establishes when the output is triggered.

`uint32_t OCPolarity;` —Define the output channel polarity when the CCRx registers matches with the CNT. This parameter can be one of the two values of the following defines:

```
TIM_OCPOLARITY_HIGH
TIM_OCPOLARITY_LOW
```

`uint32_t OCNPolarity;` —Define the complementary output polarity. It is a mode available only in TIM1 and TIM8 advanced timers, which allow two additional dedicated channels to generate complementary signals, i.e., when the channel 1 is HIGH then channel 1N is LOW and vice versa. This feature is especially useful for motor control applications. This parameter can be one of the two values of the following defines:

```
TIM_OCNPOLARITY_HIGH
TIM_OCNPOLARITY_LOW
```

`uint32_t OCFastMode;` —Specifies the fast mode state. This parameter is valid only in PWM1 and PWM2 mode. This parameter can be one of the two values of the following defines:

```
TIM_OCFAST_DISABLE
TIM_OCFAST_ENABLE
```

`uint32_t OCIdleState;` —Specifies the state of the channel output compare pin during the timer idle state. This parameter can be one of the two values of the following defines:

```
TIM_OCIDLESTATE_SET
TIM_OCIDLESTATE_RESET
```

NOTE   *This parameter is available only in TIM1 and TIM8 advanced timers.*

`uint32_t OCNIdleState;`—Specifies the state of the complimentary channel output compare pin during the timer idle state. This parameter can be one of the two values of the following defines:

```
TIM_OCNIDLESTATE_SET
TIM_OCNIDLESTATE_RESET
```

NOTE *This parameter is available only in TIM1 and TIM8 advanced timers.*

Enabling the PWM Function

You now need to enable the PWM alternative function in order to have the pulse width modulation function. The timer TIM2 is initially configured when creating the new project and having the Pinout window displayed. Figure 10-4 shows how TIM2 peripheral with PWM functionality is enabled in the Pinout view by selecting PWM direct output for channel 1 in TIM2. The PA1 pin is the channel 1 output.

The CubeMX application will automatically insert an initialization method `static void MX_TIM2_Init(void)` in the main.c file when the TIM2 PWM function is enabled. Some additional parameters should also be set in the TIM2 object using the CubeMX Configuration tab, as shown in Figure 10-5.

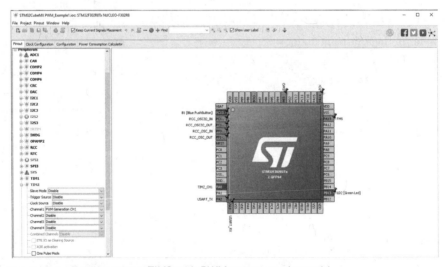

Figure 10-4 *Enabling timer TIM2 with PWM output on channel 1.*

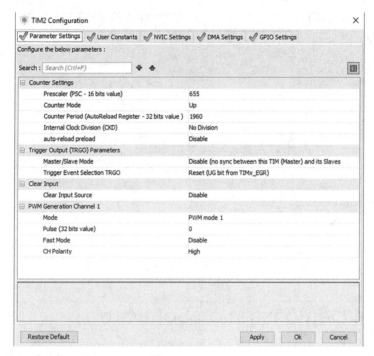

Figure 10-5 *Configuration menu settings.*

All of the initial PWM parameters are automatically set for you in the C code based upon the parameters you set in the TIM2 PWM IP tree view and the Configuration menu as shown in the following code snippet:

```
static void MX_TIM2_Init(void)
{
  TIM_MasterConfigTypeDef sMasterConfig;
  TIM_OC_InitTypeDef sConfigOC;

  htim2.Instance = TIM2;
  htim2.Init.Prescaler = 655;
  htim2.Init.CounterMode = TIM_COUNTERMODE_UP;
  htim2.Init.Period = 1960;
  htim2.Init.ClockDivision = TIM_CLOCKDIVISION_DIV1;
  htim2.Init.AutoReloadPreload = TIM_AUTORELOAD_PRELOAD_DISABLE;
  if (HAL_TIM_PWM_Init(&htim2) != HAL_OK)
  {
    _Error_Handler(__FILE__, __LINE__);
  }

  sMasterConfig.MasterOutputTrigger = TIM_TRGO_RESET;
```

```
    sMasterConfig.MasterSlaveMode = TIM_MASTERSLAVEMODE_DISABLE;
    if (HAL_TIMEx_MasterConfigSynchronization(&htim2, &sMasterConfig) !=
HAL_OK)
    {
       _Error_Handler(__FILE__, __LINE__);
    }

    sConfigOC.OCMode = TIM_OCMODE_PWM1;
    sConfigOC.Pulse = 0;
    sConfigOC.OCPolarity = TIM_OCPOLARITY_HIGH;
    sConfigOC.OCFastMode = TIM_OCFAST_DISABLE;
    if (HAL_TIM_PWM_ConfigChannel(&htim2, &sConfigOC, TIM_CHANNEL_1) !=
HAL_OK)
    {
       _Error_Handler(__FILE__, __LINE__);
    }

    HAL_TIM_MspPostInit(&htim2);

}
```

PWM Demonstration Software

The newly generated main.c file now has a method that configures and initializes the TIM2 internal peripheral with a PWM output on its channel 1. There is no functional code running within the forever loop. The PWM signal generation is all done automatically in the background. This code is shown in the following code listing:

```
// STM disclaimer goes here
// Includes
#include "main.h"
#include "stm32f3xx_hal.h"

/* USER CODE BEGIN Includes */

/* USER CODE END Includes */

// Private variables
TIM_HandleTypeDef htim2;

/* USER CODE BEGIN PV */
// Private variables

/* USER CODE END PV */

// Private function prototypes
void SystemClock_Config(void);
```

```
static void MX_GPIO_Init(void);
static void MX_TIM2_Init(void);
void HAL_TIM_MspPostInit(TIM_HandleTypeDef *htim);


/* USER CODE BEGIN PFP */
// Private function prototypes

/* USER CODE END PFP */

/* USER CODE BEGIN 0 */

/* USER CODE END 0 */

int main(void)
{

  /* USER CODE BEGIN 1 */

  /* USER CODE END 1 */

  // MCU Configuration

  // Reset all peripherals, initialize the flash and Systick
  HAL_Init();

  /* USER CODE BEGIN Init */

  /* USER CODE END Init */

  /* Configure the system clock */
  SystemClock_Config();

  /* USER CODE BEGIN SysInit */

  /* USER CODE END SysInit */

  /* Initialize all configured peripherals */
  MX_GPIO_Init();
  MX_TIM2_Init();

  /* USER CODE BEGIN 2 */
  HAL_TIM_PWM_Start(&htim2, TIM_CHANNEL_1);

      TIM2->CCR1 = 980; // CCR1 register set for 50% duty cycle
  /* USER CODE END 2 */

  /* Infinite loop */
  /* USER CODE BEGIN WHILE */
  while (1)
  {
```

```
  /* USER CODE END WHILE */

  /* USER CODE BEGIN 3 */

  }
  /* USER CODE END 3 */

}

// System Clock Configuration
void SystemClock_Config(void)
{

  RCC_OscInitTypeDef RCC_OscInitStruct;
  RCC_ClkInitTypeDef RCC_ClkInitStruct;

  // Initializes the CPU, AHB and APB buss clocks
  RCC_OscInitStruct.OscillatorType = RCC_OSCILLATORTYPE_HSI;
  RCC_OscInitStruct.HSIState = RCC_HSI_ON;
  RCC_OscInitStruct.HSICalibrationValue = 16;
  RCC_OscInitStruct.PLL.PLLState = RCC_PLL_ON;
  RCC_OscInitStruct.PLL.PLLSource = RCC_PLLSOURCE_HSI;
  RCC_OscInitStruct.PLL.PLLMUL = RCC_PLL_MUL16;
  if (HAL_RCC_OscConfig(&RCC_OscInitStruct) != HAL_OK)
  {
    _Error_Handler(__FILE__, __LINE__);
  }

  // Initializes the CPU, AHB and APB buss clocks
  RCC_ClkInitStruct.ClockType =
  RCC_CLOCKTYPE_HCLK|RCC_CLOCKTYPE_SYSCLK
  |RCC_CLOCKTYPE_PCLK1|RCC_CLOCKTYPE_PCLK2;
  RCC_ClkInitStruct.SYSCLKSource = RCC_SYSCLKSOURCE_PLLCLK;
  RCC_ClkInitStruct.AHBCLKDivider = RCC_SYSCLK_DIV1;
  RCC_ClkInitStruct.APB1CLKDivider = RCC_HCLK_DIV2;
  RCC_ClkInitStruct.APB2CLKDivider = RCC_HCLK_DIV1;

  if (HAL_RCC_ClockConfig(&RCC_ClkInitStruct, FLASH_LATENCY_2) !=
HAL_OK)
  {
    _Error_Handler(__FILE__, __LINE__);
  }

  // Configure the Systick interrupt time
  HAL_SYSTICK_Config(HAL_RCC_GetHCLKFreq()/1000);

  // Configure the Systick
  HAL_SYSTICK_CLKSourceConfig(SYSTICK_CLKSOURCE_HCLK);

  /* SysTick_IRQn interrupt configuration */
  HAL_NVIC_SetPriority(SysTick_IRQn, 0, 0);
```

```
}

/* TIM2 init function */
static void MX_TIM2_Init(void)
{

  TIM_MasterConfigTypeDef sMasterConfig;
  TIM_OC_InitTypeDef sConfigOC;

  htim2.Instance = TIM2;
  htim2.Init.Prescaler = 655;
  htim2.Init.CounterMode = TIM_COUNTERMODE_UP;
  htim2.Init.Period = 1960;
  htim2.Init.ClockDivision = TIM_CLOCKDIVISION_DIV1;
  htim2.Init.AutoReloadPreload = TIM_AUTORELOAD_PRELOAD_DISABLE;
  if (HAL_TIM_PWM_Init(&htim2) != HAL_OK)
  {
     _Error_Handler(__FILE__, __LINE__);
  }

  sMasterConfig.MasterOutputTrigger = TIM_TRGO_RESET;
  sMasterConfig.MasterSlaveMode = TIM_MASTERSLAVEMODE_DISABLE;
  if (HAL_TIMEx_MasterConfigSynchronization(&htim2, &sMasterConfig) !=
HAL_OK)
  {
     _Error_Handler(__FILE__, __LINE__);
  }

  sConfigOC.OCMode = TIM_OCMODE_PWM1;
  sConfigOC.Pulse = 0;
  sConfigOC.OCPolarity = TIM_OCPOLARITY_HIGH;
  sConfigOC.OCFastMode = TIM_OCFAST_DISABLE;
  if (HAL_TIM_PWM_ConfigChannel(&htim2, &sConfigOC, TIM_CHANNEL_1) !=
HAL_OK)
  {
     _Error_Handler(__FILE__, __LINE__);
  }

  HAL_TIM_MspPostInit(&htim2);

}

/** Configure pins as
        * Analog
        * Input
        * Output
        * EVENT_OUT
        * EXTI
     PA2    ------> USART2_TX
     PA3    ------> USART2_RX
*/
```

```
static void MX_GPIO_Init(void)
{

  GPIO_InitTypeDef GPIO_InitStruct;

  /* GPIO Ports Clock Enable */
  __HAL_RCC_GPIOC_CLK_ENABLE();
  __HAL_RCC_GPIOF_CLK_ENABLE();
  __HAL_RCC_GPIOA_CLK_ENABLE();
  __HAL_RCC_GPIOB_CLK_ENABLE();

  /*Configure GPIO pin Output Level */
  HAL_GPIO_WritePin(LD2_GPIO_Port, LD2_Pin, GPIO_PIN_RESET);

  /*Configure GPIO pin : B1_Pin */
  GPIO_InitStruct.Pin = B1_Pin;
  GPIO_InitStruct.Mode = GPIO_MODE_IT_FALLING;
  GPIO_InitStruct.Pull = GPIO_NOPULL;
  HAL_GPIO_Init(B1_GPIO_Port, &GPIO_InitStruct);

  /*Configure GPIO pins : USART_TX_Pin USART_RX_Pin */
  GPIO_InitStruct.Pin = USART_TX_Pin|USART_RX_Pin;
  GPIO_InitStruct.Mode = GPIO_MODE_AF_PP;
  GPIO_InitStruct.Pull = GPIO_NOPULL;
  GPIO_InitStruct.Speed = GPIO_SPEED_FREQ_LOW;
  GPIO_InitStruct.Alternate = GPIO_AF7_USART2;
  HAL_GPIO_Init(GPIOA, &GPIO_InitStruct);

  /*Configure GPIO pin : LD2_Pin */
  GPIO_InitStruct.Pin = LD2_Pin;
  GPIO_InitStruct.Mode = GPIO_MODE_OUTPUT_PP;
  GPIO_InitStruct.Pull = GPIO_NOPULL;
  GPIO_InitStruct.Speed = GPIO_SPEED_FREQ_LOW;
  HAL_GPIO_Init(LD2_GPIO_Port, &GPIO_InitStruct);

}

/* USER CODE BEGIN 4 */

/* USER CODE END 4 */

/**
  * @brief  This function is executed in case of error occurrence.
  * @param  None
  * @retval None
  */
void _Error_Handler(char * file, int line)
{
  /* USER CODE BEGIN Error_Handler_Debug */
  /* User can add his own implementation to report the HAL error return
state */
```

```
    while(1)
    {
    }
    /* USER CODE END Error_Handler_Debug */
}

#ifdef USE_FULL_ASSERT

/**
   * @brief Reports the name of the source file and the source line number
   * where the assert_param error has occurred.
   * @param file: pointer to the source file name
   * @param line: assert_param error line source number
   * @retval None
   */
void assert_failed(uint8_t* file, uint32_t line)
{
  /* USER CODE BEGIN 6 */
  /* User can add his own implementation to report the file name and line number,
     ex: printf("Wrong parameters value: file %s on line %d\r\n", file,
line) */
  /* USER CODE END 6 */

}

#endif

/**
  * @}
  */

/**
  * @}
  */

/**** (C) COPYRIGHT STMicroelectronics ****/
```

You should note that I set the prescaler to a 655 value. This meant that the TIM2 clock input would be at a frequency approximately equal to 977 kHz with a period of 10.23 µs. This value ensured that the timer resolution would allow for setting very precise pulse widths.

Demonstration One

This demonstration uses the values set during the initial configuration to generate a 50% duty cycle waveform. The program was first built and then downloaded into the project board. I used a USB oscilloscope to observe the PWM signal

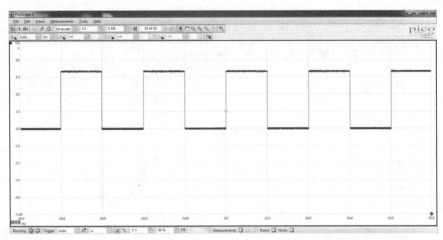

Figure 10-6 *Fifty percent duty-cycle PWM waveform.*

generated from the TIM2, channel 1 output, which is emitted from pin PA1. Figure 10-6 shows this waveform.

You should be able to see that the period of the over signal is 20 ms and the high time for the pulse is 10 ms, which creates the 50% duty cycle as expected. This waveform used a prescaler value equal to 655, which caused each timer count to be 10.23 μs as discussed above. The Period value was set at 1960, which means the actual period time is 1960 times 10.23 μs or 20050.8 μs, equivalent to 20.0508 ms. That was close enough for my purposes. The CCR1 register held a value of 980, which meant the output switched at 980 times 10.23 μs or 10025.4 μs, equivalent to 10.0254 ms. This is extremely close to a 50% duty cycle for all practical purposes. This demonstration proved that the PWM generation was working as desired.

Table 10-1 is provided as a handy reference showing a series of CCRx values that will generate a respective duty cycle given that value. Note that the small duty cycles shown at the table's beginning are important for the servo control demonstration. This table is based upon a 50 Hz or 20 ms period PWM signal.

Demonstration Two

This demonstration controls the intensity or luminescence of a single, red LED, using a PWM output as shown in the previous demonstration. The only parameter to be varied is the CCRx value assigned in the main method by this statement:

```
TIM2->CCR1 = 980; // CCR1 register set for 50% duty cycle
```

CCRx (value)	Duty Cycle (%)	Pulse High Time (ms)
98	5	1
148	7.5	1.5
196	10	2
392	20	4
588	30	6
784	40	8
980	50	10
1176	60	12
1372	70	14
1568	80	16
1764	90	18
1960	100	20

Table 10-1   *CCRx Values and Corresponding Duty Cycles and Pulse High Times*

I chose to vary this value by editing the main.c file and then quickly rebuilding it. This process was very fast and not cumbersome at all. You could also set up an OpenOCD session and adjust the value in a dynamic fashion, but I found the edit, build, and download process to be very convenient without the bother of creating a telnet session. I would suggest you try both approaches and use whichever you find useful and productive.

Connecting Channel 1 PWM Output to an LED/Resistor

Figure 10-7 shows the connections for the same LED/resistor combination, which I earlier used for the project demonstration in Chapter 5. The LED/resistor is connected between PA1 and ground using the Arduino protoboard.

Test Results

I observed that the LED luminescence did vary considerably with the duty cycle of the applied PWM signal. You should realize one important distinction between dimming an incandescent bulb and an LED. The incandescent bulb intensity corresponds to the average voltage applied to it, while the LED luminescence depends on your eye's integrative response. By this, I mean the voltage applied to the LED is at a constant level, but the time interval applied varies with the duty cycle. The human eye will tend to integrate or average out the LED light intensity and thus perceive different intensities for varying duty cycles. The eye also has a nonlinear relationship between actual luminance and perceived brightness, as shown in Figure 10-8.

Figure 10-7 *LED/resistor connected on the Arduino protoboard.*

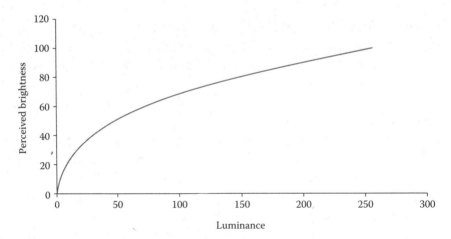

Figure 10-8 *Human eye brightness perception.*

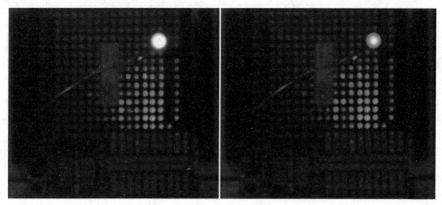

Figure 10-9   *100% and 10% PWM signal-duty cycles applied to a red LED.*

What this all means is that the eye will more easily recognize different LED luminescence at small duty cycles but will not perceive any significant differences as the duty cycle increases. There are equations that model the eye brightness to luminance relationship quite well and have been codified as the CIE1931 lightness formula. White light LED manufacturers have modified their products using this relationship to make their commercial LEDs have a linear dimming property. I have not done this with this single LED demonstration, but it is quite doable and you would put the code into the main.c forever loop.

I did try to capture the intensity difference between 100% and 10% duty cycles. Figure 10-9 is a composite photograph with the 100% PWM on the left and the 10% PWM on the right.

The camera shutter speed that was required for these images was fairly long, which tended to make the 10% image much brighter than it actually was. I guess you will just have to duplicate this demonstration and see the actual duty cycle versus perceived luminance for yourself.

Demonstration Three

In this demonstration I will be controlling a RGB LED that has three input lines, one for each color. GP timer TIM2 will be reconfigured to have three PWM outputs connected to each of the LED's lines. The PWM lines will control both the LED color and luminescence based on the duty cycle set on each line.

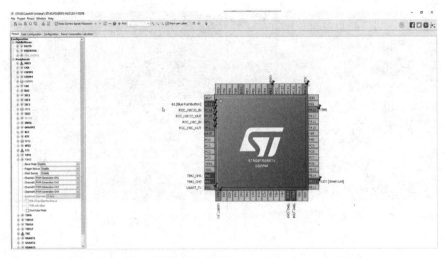

Figure 10-10   *Enabling timer TIM2 with PWM output on channels 1 to 4.*

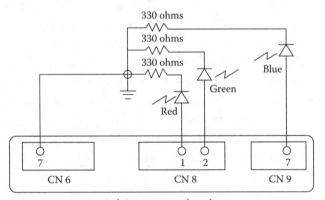

Figure 10-11   *RGB to Arduino protoboard connection schematic.*

The first step is to create a new CubeMX project to handle the additional PWM lines. Figure 10-10 shows the Pinout view where I have configured TIM2 for four channel outputs, of which I will use the first three to control the RGB LED.

The configuration menu is exactly the same as previously shown in Figure 10-5.

Connecting Channel PWM Outputs to a RGB LED

Figure 10-11 is a schematic showing the connections between the RGB LED and the timer GPIO pins as connected on the Arduino protoboard. Note that each

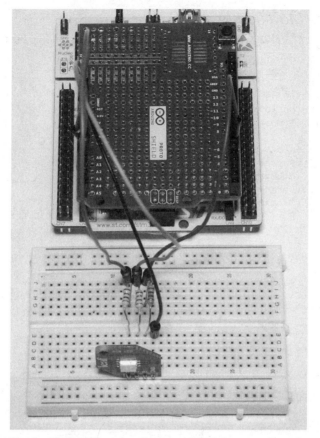

Figure 10-12 *Physical RGB LED connections to an Arduino protoboard.*

LED line has a series current limiting resistor to protect both the GPIO pin and the RGB LED.

The timer TIM2 channel outputs are on the following pins:

- Channel 1—PA0
- Channel 2—PA1
- Channel 3—PB10
- Channel 4—PB11

Figure 10-12 shows the physical setup for a RGB LED connected to an Arduino protoboard.

Software Modifications

A modified `MX_TIM2_Init` method will be added to the main.c file based on the additional channel outputs. This new method is listed below:

```
static void MX_TIM2_Init(void)
{
  TIM_MasterConfigTypeDef sMasterConfig;
  TIM_OC_InitTypeDef sConfigOC;

  htim2.Instance = TIM2;
  htim2.Init.Prescaler = 655;
  htim2.Init.CounterMode = TIM_COUNTERMODE_UP;
  htim2.Init.Period = 1960;
  htim2.Init.ClockDivision = TIM_CLOCKDIVISION_DIV1;
  htim2.Init.AutoReloadPreload = TIM_AUTORELOAD_PRELOAD_DISABLE;
  if (HAL_TIM_PWM_Init(&htim2) != HAL_OK)
  {
    _Error_Handler(__FILE__, __LINE__);
  }
  sMasterConfig.MasterOutputTrigger = TIM_TRGO_RESET;
  sMasterConfig.MasterSlaveMode = TIM_MASTERSLAVEMODE_DISABLE;
  if (HAL_TIMEx_MasterConfigSynchronization(&htim2, &sMasterConfig) !=
HAL_OK)
  {
    _Error_Handler(__FILE__, __LINE__);
  }

  sConfigOC.OCMode = TIM_OCMODE_PWM1;
  sConfigOC.Pulse = 0;
  sConfigOC.OCPolarity = TIM_OCPOLARITY_HIGH;
  sConfigOC.OCFastMode = TIM_OCFAST_DISABLE;
  if (HAL_TIM_PWM_ConfigChannel(&htim2, &sConfigOC, TIM_CHANNEL_1) !=
HAL_OK)
  {
    _Error_Handler(__FILE__, __LINE__);
  }
  if (HAL_TIM_PWM_ConfigChannel(&htim2, &sConfigOC, TIM_CHANNEL_2) !=
HAL_OK)
  {
    _Error_Handler(__FILE__, __LINE__);
  }
  if (HAL_TIM_PWM_ConfigChannel(&htim2, &sConfigOC, TIM_CHANNEL_3) !=
HAL_OK)
  {
    _Error_Handler(__FILE__, __LINE__);
  }
  if (HAL_TIM_PWM_ConfigChannel(&htim2, &sConfigOC, TIM_CHANNEL_4) !=
HAL_OK)
  {
    _Error_Handler(__FILE__, __LINE__);
  }
```

```
HAL_TIM_MspPostInit(&htim2);
}
```

There is also some new code that must be entered into the main method. This new code is listed below:

```
/* USER CODE BEGIN 2 */

HAL_TIM_PWM_Start(&htim2, TIM_CHANNEL_1);
HAL_TIM_PWM_Start(&htim2, TIM_CHANNEL_2);
HAL_TIM_PWM_Start(&htim2, TIM_CHANNEL_3);

TIM2->CCR1 =   980; // Red channel
TIM2->CCR2 =   980; // Green channel
TIM2->CCR3 =   980; // Blue channel

/* USER CODE END 2 */
```

Note that each of the RGB channels was initialized at a 50% duty cycle.

Test Results

The RGB LED initially showed a bright white light, which was expected because all the individual LED components had the same duty cycle. I then reset the individual CCRx registers to separately test each color and confirmed that they were operating properly. I also tried varying the duty cycles to create different colors with little success because of the way the RGB LED was constructed. It did not merge the light from each colored LED and therefore did not properly create an alternate color, which you can do if you were mixing paint. I believe I would have been successful if I had used a more expensive RGB LED with some built-in optics, but unfortunately all I had was a very inexpensive RGB LED. In any case, the objective for this demonstration was to show you how to create and use multiple PWM outputs, which I believe was achieved.

Demonstration Four

This last demonstration will focus on using a single PWM signal to control a hobby-grade servo motor. I will be using the same program that was used for the first two demonstrations and no code modifications are required. I will start the demonstration by reviewing some basic facts regarding how analog servos are controlled.

Analog Servo Control

An analog servo is essentially an electric motor that has a built-in electronic circuit that converts specific pulse widths into a proportional rotation. The specifications

Pulse Width (50 Hz) (ms)	Servo Rotation with Respect to the Neutral Position (1.5 ms) (degrees)
1.0	−60
1.5 (neutral)	0
2.0	+60

Table 10-2 *Pulse Width and Servo Rotation*

Figure 10-13 *Hitec model HS-311 analog servo.*

for hobby-grade analog servos are fairly "relaxed" meaning that there are loose tolerances relating pulse width to physical rotation. This is mainly due to the use of very low cost components in both the electrical and mechanical components that comprise the servo. Table 10-2 details pulse width and expected rotation.

An analog servo should not move from its neutral position when it is sent a repetitive 1.5-ms pulse signal. A 1.0-ms pulse train will cause it to rotate its main shaft counter-clockwise 60° if viewed from head-on. Similarly, a 2.0-ms pulse train will cause it to rotate clockwise 60°. However, loose tolerances come into play and often the degree of rotation can be 10° to 15°, more or less.

In all cases, the pulse train frequency is set at a nominal 50 Hz or 20 ms period.

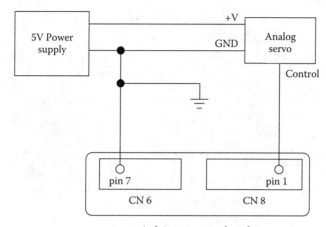

Figure 10-14 *Servo connection schematic.*

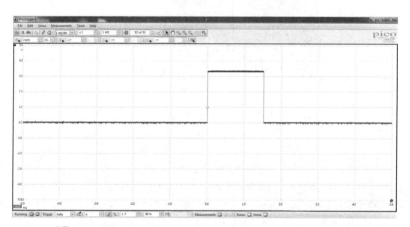

Figure 10-15 *1.5-ms servo control signal.*

Figure 10-13 shows the hobby-grade servo I used for this demonstration. It is a Hitec model HS-311, which is a reliable, well-built servo.

This servo takes quite a bit of current to operate, somewhat in excess of what the STM project board can provide. Consequently, I used a separate supply for it as may be seen in the Figure 10-14 connection schematic.

The servo control signal from the PWM output only requires several milliamperes, which is well within the GPIO pin drive capabilities.

Figure 10-15 shows the waveform for the 1.5-ms neutral position signal. It is a nice, clean signal with sharp rise and fall times, which are important for proper servo control.

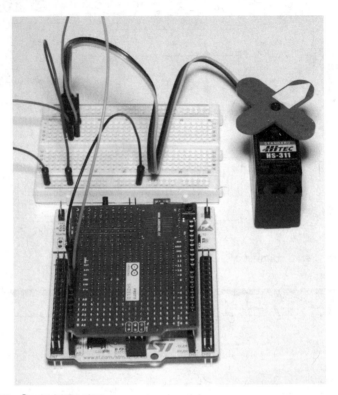

Figure 10-16 *Servo test setup.*

The 1.0-ms and 2.0-ms waveforms are very similar except for the pulse width.

Figure 10-16 shows the physical setup with the analog servo connected to a solderless breadboard, which in turn is connected to the Arduino protoboard with jumper wires.

Test Results

I first changed the CCR1 value to 148, which causes a 1.5-ms pulse width to be generated. The servo immediately moved to its neutral position after the program was downloaded into the project board. I repeated the previous step with a CCR1 equal to 98, which generates a 1.0-ms pulse width. The servo responded by moving its main shaft 60° counter-clockwise. Finally, I changed CCR1 to 196, which generates a 2.0-ms pulse width. The servo's main shaft rotated clockwise 120° as expected.

The servo performed flawlessly and showed no signs of jitter or vibration, which sometimes occurs with using hobby-grade servos.

I next added some functional code and placed it in the forever loop to continuously exercise the servo, thus eliminating the need to edit, rebuild, and download the controlling program.

Adding Functional Test Code

The following C code was placed within the main.c file in order to continuously control the servo. This code caused the servo main shaft to slowly oscillate between the ±60° end points.

```
* USER CODE BEGIN PV */
// Private variables
int pw;
/* USER CODE END PV */


/* Infinite loop */
  /* USER CODE BEGIN WHILE */
  while (1)
  {
  /* USER CODE END WHILE */
    pw = 98;
    while(pw < 197)
    {
        TIM2->CCR1 = pw;
        HAL_Delay(20);
          pw++;
    }

    while(pw > 97)
    {
        TIM2->CCR1 = pw;
        HAL_Delay(20);
        pw--;
    }
  /* USER CODE BEGIN 3 */
  }
  /* USER CODE END 3 */
```

Test Results

I observed the servo main shaft to slowly oscillate between the end points as expected after the modified code was downloaded into the project board.

This final demonstration concludes the chapter. I believe I have provided you with a sufficient background that will enable you to successfully use servos with a STM Nucleo board.

Summary

I started this chapter on pulse width modulation (PWM) by explaining what makes up a PWM signal and detailing certain key properties of that signal. This was followed by a brief discussion of how the STM timer hardware generates a PWM signal.

Next came a discussion on how the HAL framework is used to both configure and initialize a general purpose (GP) timer to emit a PWM signal.

A complete C code listing was shown that generated a PWM signal with a 50% duty cycle and a 50-Hz output frequency.

Next followed the first of four PWM demonstrations. The first PWM demonstration just output the 50% duty cycle waveform that was generated by the project board after the code was built and downloaded into the board.

The second demonstration used a varying duty-cycle waveform to dim and brighten a red LED.

The third demonstration used multi-channel PWM waveforms to control both the color and intensity of a RGB LED.

The fourth and final demonstration illustrated how to control a standard analog servo using PWM signals.

11

Direct Memory Access (DMA) and the Digital-to-Analog Converter (DAC)

This chapter content is carefully chosen to explain direct memory access (DMA) and the very useful digital-to-analog converter (DAC) peripheral, which takes advantage of efficient DMA data transfers. There are also several demonstration projects shown in the chapter, which show both DMA and DAC operations. The last one shows how to use DMA with the DAC peripheral

DMA

Data is normally transferred by using instructions executed by the Cortex-M core processor. This data is normally sent to or from internal memories and peripherals. This data transfer process consumes valuable processor cycles that could otherwise be put to use within a constrained embedded system. This is the reason why the DMA controller was designed. It can assume the data transfer process in many cases and allow the core processor to take on other critical embedded system processes.

Technically, the DMA controller can be considered a peripheral device; however, I prefer to think of it as more like a co-processor than a peripheral. Unlike any other peripheral, the DMA controller can take on the role of being a bus master, which is not allowed for any other device other than the core processor itself. In fact, thinking of the DMA controller as more of a bus controller is a good way to view this device.

241

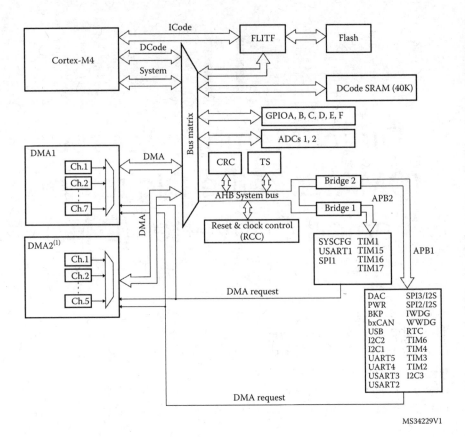

MS34229V1

Figure 11-1 *DMA block diagram.*

Figure 11-1 is a block diagram for the DMA controller used in the project board's STM32F302R8 MCU. This figure will be very useful in explaining how the DMA controller works.

NOTE *DMA2, SPI1, TIM3, TIM4, UART4, UART5, and ADC2 are not available in the STM32F302R8 MCU.*

It would be worthwhile to consider some of the basics of data transfers before going into the specifics of DMA operations.

Basic Data Transfer Concepts

Figure 11-2 is a simplified system block diagram where all the DMA components have been removed from Figure 11-1. The purpose of this diagram is to focus on

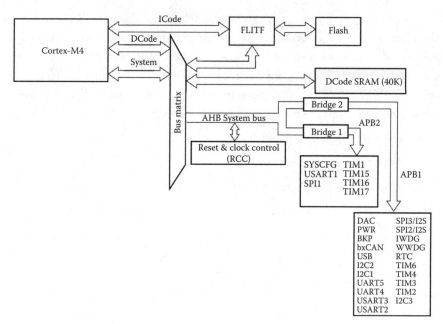

Figure 11-2 *Data transfers between the core processor and system peripherals.*

how data transfers happen between the core processor and all remaining system peripherals.

You should clearly see that all data transfers between the core Cortex-M processor and any peripheral or internal memory must go through either the Bus Matrix or FLITF blocks. There is simply no other way that data can reach the processor. Every data transfer requires some amount of processor cycles to complete. For example, the processor must handle all data sent by an external device connected to a UART peripheral where that data is to be designated to be stored in SRAM. It is important to realize that the processor does not modify or change the data in any way, but simply sends it along to the proper memory storage locations. This requirement for core processor involvement ties up precious cycles that otherwise could be utilized for more efficient operations. A somewhat analogous model is where data is represented by cars and the processor by a city. Cross-country drivers wanting to go to a distant location are forced to go through a city with all its traffic lights and congested streets. It would be much better for the drivers if a bypass highway were available, allowing them to avoid going through the city. The DMA controller plays the role of the bypass.

Returning to Figure 11-1, you should be able to see that DMA1 controller, in the case of the STM32F302R8 MCU, connects directly to the Bus Matrix. This controller, as I earlier mentioned, is set up to be a bus controller or master and thus is able to configure the Bus Matrix to directly connect data between peripherals and internal memory, memory to memory, or even to another peripheral. These configurations are naturally named as follows:

- Peripheral-to-memory transfers

- Memory-to-peripheral transfers

- Peripheral-to-peripheral transfers

- Memory-to-memory transfers

The Bus Matrix controls the access of both the core processor and the DMA controller to various data busses, memory, and peripherals. It implements this control with a Round Robin algorithm, which ensures equitable access to both controllers. Figure 11-3 is a block diagram illustrating how the Bus Matrix is connected with all the system components.

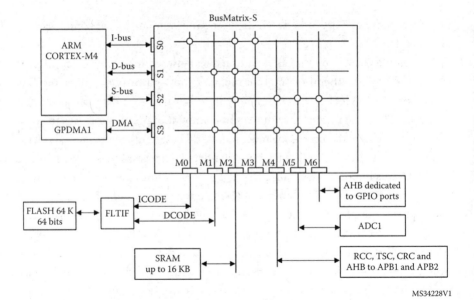

MS34228V1

Figure 11-3 *Bus Matrix block diagram.*

The Bus Matrix is a complex circuit that interconnects the following master and slave devices:

- Core Cortex-M processor (master)

- DMA controller(s) (master)

- Flash memory interface (FLITF) using the ICODE (slave)

- Flash memory interface (FLITF) using the DCODE buss (slave)

- Direct connection to SRAM (slave)

- Advanced High-Speed Bus 1 (AHB1), which connects to the Advanced Peripheral Busses (AHB1 and AHP2) through bridge circuits (Bridge 1 and Bridge 2) (slave)

- Advanced Peripheral Bus (APB) connected to the GPIO pins (slave)

- Direct connection to the ADC1 peripheral (slave)

DMA Controller Details

Every STM32 DMA controller contains two master ports, one named the memory port, which connects to internal memory such as SRAM and flash. The other is named the peripheral port and it connects to the Advanced High-Performance Bus (AHB). The AHB in turn connects with the slave peripherals. In the STM32F302R8 DMA controller, the peripheral port can also connect to a memory control circuit permitting memory-to-memory transfers.

STM32 DMA controllers also have one slave port connected to the AHB bus. This allows the DMA controller to be programmed from the other master, that is, the core processor.

The STM32F302R8 DMA controller has seven independent and programmable channels, which are each connected to separate hardware lines. These are how the slave peripherals can request a response from the DMA controller. The exact nature and type of request is configured during the MCU hardware design phase. Different STM MCUs will have a varying numbers of channels as well as different request types depending upon the MCU design requirements.

There is also a priority assignment hardware unit, which allows different priorities to be assigned to the seven DMA request channels. Higher priorities will normally be assigned to fast peripherals needing immediate DMA response as well as to any memory accesses.

Figure 11-4 shows all seven DMA channels along with their respective peripheral request signals that are bound to them during the MCU design phase.

Figure 11-4 is an important figure because it will be your ready reference to use when trying to determine which DMA channel to use for a given peripheral DMA request. It is very important to note that only one peripheral can be active on a given peripheral request line.

Notice in Figure 11-4 that the DMA channels have default priorities, which are fixed in hardware with channel 1 being the highest and progressing to channel 7 as the lowest. However, these default priorities may be overridden with user-assigned priorities, designed to determine which peripherals have preference when accessing the AHB. DMA channel lines are normally trigged through hardware signals from selected peripherals but they can also be activated by software. Software triggering is the means by which memory-to-memory transfers are activated.

Using HAL with DMA

The HAL framework uses a C struct named `DMA_HandleTypeDef` to configure the DMA controller. Each member will separately be discussed after the struct definition.

```
typedef struct {
DMA_Channel_TypeDef          *Instance;
DMA_InitTypeDef              Init;
HAL_LockTypeDef              Lock;
__IO HAL_DMA_StateTypeDef    State;
void                         *Parent;
void            (* XferCpltCallback)(struct __DMA_HandleTypeDef *hdma);
void            (* XferHalfCpltCallback)(struct __DMA_HandleTypeDef *hdma);
void            (* XferErrorCallback)(struct __DMA_HandleTypeDef *hdma);
__IO uint32_t                ErrorCode;
} DMA_HandleTypeDef;
```

The `DMA_HandleTypeDef` struct members are briefly defined in the following list:

- `DMA_Channel_TypeDef *Instance;` —This is the pointer to the DMA/Channel pair descriptor which will be used. For example, DMA1_Channel5 indicates the fifth channel of DMA1. Recall that these designations are bound to peripherals during the MCU design. Refer to Figure 11-4 to see the channel-to-peripheral designations.

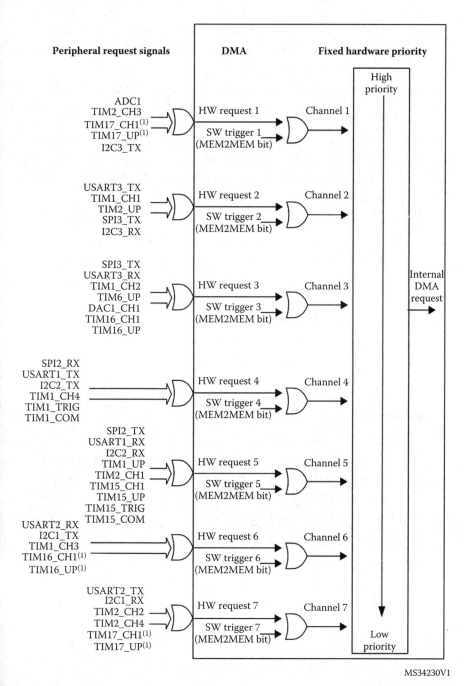

MS34230V1

Figure 11-4 *DMA channel block diagram.*

- `DMA_InitTypeDef Init;`—This is an instance of the C struct `DMA_InitTypeDef`, which is used to configure the DMA/Channel pair. This value will be further explained below.

- `HAL_LockTypeDef Lock;`—DMA locking object.

- `__IO HAL_DMA_StateTypeDef State;`—DMA transfer state.

- `void *Parent;`—This is a pointer used to keep track of the peripheral handlers associated with the current DMA/Channel. For example, if we are using a UART in DMA mode, this field will point to an instance of `UART_HandleTypeDef`.

- `void (* XferCpltCallback)(struct __DMA_HandleTypeDef *hdma);`—This is a pointer to a callback function, which is used to signal user code that a DMA transfer is completed. It is automatically called by the HAL when a DMA interrupt is paired, by the function `HAL_DMA_IRQHandler()`.

- `void (* XferHalfCpltCallback)(struct __DMA_HandleTypeDef *hdma);`—This is a pointer to a callback function, which is used to signal user code that a DMA transfer is completed, half-completed, or an error occurred. It is automatically called by the HAL when a DMA interrupt is paired by the function `HAL_DMA_IRQHandler()`.

- `void (* XferErrorCallback)(struct __DMA_HandleTypeDef *hdma);`—This is a pointer to a callback function, which is used to signal user code that a DMA transfer has an error. It is automatically called by the HAL when a DMA interrupt is paired by the function `HAL_DMA_IRQHandler()`.

- `__IO uint32_t ErrorCode;`—DMA error code.

The HAL framework uses a C struct named `DMA_InitTypeDef` to initialize the DMA controller. Each member will separately be discussed after the struct definition.

```
typedef struct {
uint32_t Direction;
uint32_t PeriphInc;
uint32_t MemInc;
uint32_t PeriphDataAlignment;
uint32_t MemDataAlignment;
```

```
uint32_t Mode;
uint32_t Priority;
} DMA_InitTypeDef;
```

The `DMA_InitTypeDef` struct members are briefly defined in the following list:

- `uint32_t Direction;`—This value defines the DMA transfer direction. This parameter can be any value of the following defines:

  - `DMA_PERIPH_TO_MEMORY`

  - `DMA_MEMORY_TO_PERIPH`

  - `DMA_MEMORY_TO_MEMORY`

- `uint32_t PeriphInc;`—This value can automatically increment the peripheral register holding the data address to be transferred. This parameter can be one of the two values of the following defines:

  - `DMA_PINC_ENABLE`

  - `DMA_PINC_DISABLE`

- `uint32_t MemInc;`—This value can automatically increment the memory port register holding the memory data address to be transferred. This parameter can be one of the two values of the following defines:

  - `DMA_MINC_ENABLE`

  - `DMA_MINC_DISABLE`

- `uint32_t PeriphDataAlignment;`—This value set the sizes of the peripheral data elements to be transferred. This parameter can be any value of the following defines:

  - `DMA_PDATAALIGN_BYTE`

  - `DMA_PDATAALIGN_HALFWORD`

  - `DMA_PDATAALIGN_WORD`

- `uint32_t MemDataAlignment;`—This value set the sizes of the memory data elements to be transferred. This parameter can be any value of the following defines:

  - `DMA_MDATAALIGN_BYTE`

  - `DMA_MDATAALIGN_HALFWORD`

  - `DMA_MDATAALIGN_WORD`

- `uint32_t Mode;`—This value sets the DMA mode. In the normal mode, the DMA controller sends the specified amount of data from the source port to the destination port and then stops. The controller must be rearmed for another transfer. In the circular mode, the DMA controller sends the content of the source data buffer, resets to the first byte in source data buffer, and starts sending the buffer again. In the transmission interim, the peripheral will have had a chance to refill the source buffer so there will be fresh data in the source buffer. In this case, the source buffer is also called a ring buffer. The DMA circular mode is also called the continuous mode for obvious reasons. This parameter can be one of the two values of the following defines:

    - DMA_NORMAL

    - DMA_CIRCULAR

- `uint32_t Priority;`—Use this value to assign a priority to a DMA channel. This parameter can be any value of the following defines:

    - DMA_PRIORITY_LOW

    - DMA_PRIORITY_MEDIUM

    - DMA_PRIORITY_HIGH

    - DMA_PRIORITY_VERY_HIGH

Final Steps in Setting Up a DMA Transfer

There are just a few more steps required in order to configure a DMA transfer over a selected channel. These are as follows:

1. Set up appropriate memory and/or peripheral port addresses

2. Establish the size of the data transfer

3. Arm the DMA controller

4. Enable the DMA mode on the selected peripheral

The HAL framework abstracts the first three steps with the following statement:

```
HAL_StatusTypeDef HAL_DMA_Start(DMA_HandleTypeDef *hdma,
uint32_t SrcAddress, uint32_t Dst\ Address, uint32_t
DataLength);
```

The fourth step is dependent on the selected peripheral. You will normally consult the STM data sheet, which describes the selected peripheral in order to determine how to enable the DMA transfer and to find out the exact channel it has been preassigned. Remember from the earlier discussion that peripherals have been preassigned channel numbers during the MCU design phase. In the following example, I will be demonstrating how to use USART2 peripheral in the DMA mode.

Demonstration One

This first demonstration will involve sending a string using the USART2 peripheral operating in the normal DMA mode. There are just a few steps necessary to configure this demonstration:

1. First configure the USART2 peripheral using the Pinout view in the CubeMX application as has been done several times in previous demonstrations.

2. Set the DMA1 channel as follows (STM32F302R8 MCU) for the USART2 peripheral for a memory-to-peripheral transfer:

   a. USART2_TX = DMA1/CH7

   b. USART2_RX = DMA1/CH6

   The following statement sets the USART2 peripheral DMA1, channel 7 to output characters to a terminal program:

   ```
   hdma_usart2_tx.Instance +DMA_Channel7;
   ```

   It must be put in the main method. In addition, there is no need to instantiate DMA1, channel 6 because no characters will be received from the terminal program in this demonstration.

3. Enable the USART2 peripheral for DMA operations and arm the appropriate channel to execute the transfer.

   The following statement enables the USART2 peripheral for DMA operations:

   ```
   HAL_DMA_Init(&hdma_usart2_tx);
   ```

   It must be put in the main method.

Program Listing

The following is the code listing for this demonstration. I have also provided some additional discussion regarding pertinent code segments following the listing.

```c
// STM disclaimer goes here
// Includes
#include "main.h"
#include "stm32f3xx_hal.h"
#include <string.h>

/* USER CODE BEGIN Includes */

/* USER CODE END Includes */

// Private variables
UART_HandleTypeDef huart2;
DMA_HandleTypeDef hdma_usart2_tx;

/* USER CODE BEGIN PV */
// Private variables
    char msg[] = "Hello World";   // This is the actual data
/* USER CODE END PV */

// Private function prototypes
void SystemClock_Config(void);
static void MX_GPIO_Init(void);
static void MX_DMA_Init(void);
static void MX_USART2_UART_Init(void);

/* USER CODE BEGIN PFP */
// Private function prototypes

/* USER CODE END PFP */

/* USER CODE BEGIN 0 */

/* USER CODE END 0 */

int main(void)
{
  // MCU Configuration

  // Reset all peripherals; initialize the flash and the Systick.
  HAL_Init();

  /* USER CODE BEGIN Init */

  /* USER CODE END Init */

  /* Configure the system clock */
  SystemClock_Config();
```

```
/* USER CODE BEGIN SysInit */

/* USER CODE END SysInit */

/* Initialize all configured peripherals */
MX_GPIO_Init();
MX_DMA_Init();
MX_USART2_UART_Init();

/* USER CODE BEGIN 2 */

/* USER CODE END 2 */
/* USER CODE BEGIN 1 */
    hdma_usart2_tx.Instance = DMA1_Channel7;   // Bound channel
        hdma_usart2_tx.Init.Direction = DMA_MEMORY_TO_PERIPH;
        hdma_usart2_tx.Init.PeriphInc = DMA_PINC_DISABLE;
        hdma_usart2_tx.Init.MemInc = DMA_MINC_ENABLE;
        hdma_usart2_tx.Init.PeriphDataAlignment = DMA_PDATAALIGN_BYTE;
        hdma_usart2_tx.Init.MemDataAlignment = DMA_MDATAALIGN_BYTE;
        hdma_usart2_tx.Init.Mode = DMA_NORMAL;
        hdma_usart2_tx.Init.Priority = DMA_PRIORITY_LOW;

    // Initialize the UART DMA operation
    HAL_DMA_Init(&hdma_usart2_tx);

    // Start the DMA transfer from peripheral to memory
    HAL_DMA_Start(&hdma_usart2_tx, (uint32_t)msg, (uint32_t)&huart2.
    Instance->TDR, (uint32_t)strlen(msg));

    // Enable USART2 to start data transfer
        huart2.Instance->CR3 |= USART_CR3_DMAT;

    // Poll for DMA operation completion
        HAL_DMA_PollForTransfer(&hdma_usart2_tx, HAL_DMA_FULL_TRANSFER,
        HAL_MAX_DELAY);

    // Toggle the USART2 CR3 register after data transfer completed
    huart2.Instance->CR3 &= ~USART_CR3_DMAT;

    // Turn on the user LED to indicate DMA operation is done!
    HAL_GPIO_WritePin(LD2_GPIO_Port, LD2_Pin, GPIO_PIN_SET);


    /* USER CODE END 1 */

/* Infinite loop */
/* USER CODE BEGIN WHILE */
while (1)
{
/* USER CODE END WHILE */
```

```
    /* USER CODE BEGIN 3 */
    // Nothing required in the forever loop
    }
    /* USER CODE END 3 */

}

// System Clock Configuration
void SystemClock_Config(void)
{
    RCC_OscInitTypeDef RCC_OscInitStruct;
    RCC_ClkInitTypeDef RCC_ClkInitStruct;

    // Initializes the CPU, AHB and APB buss clocks
    RCC_OscInitStruct.OscillatorType = RCC_OSCILLATORTYPE_HSI;
    RCC_OscInitStruct.HSIState = RCC_HSI_ON;
    RCC_OscInitStruct.HSICalibrationValue = 16;
    RCC_OscInitStruct.PLL.PLLState = RCC_PLL_ON;
    RCC_OscInitStruct.PLL.PLLSource = RCC_PLLSOURCE_HSI;
    RCC_OscInitStruct.PLL.PLLMUL = RCC_PLL_MUL16;
    if (HAL_RCC_OscConfig(&RCC_OscInitStruct) != HAL_OK)
    {
        _Error_Handler(__FILE__, __LINE__);
    }

    // Initializes the CPU, AHB and APB buss clocks
    RCC_ClkInitStruct.ClockType =
    RCC_CLOCKTYPE_HCLK|RCC_CLOCKTYPE_SYSCLK
    |RCC_CLOCKTYPE_PCLK1|RCC_CLOCKTYPE_PCLK2;
    RCC_ClkInitStruct.SYSCLKSource = RCC_SYSCLKSOURCE_PLLCLK;
    RCC_ClkInitStruct.AHBCLKDivider = RCC_SYSCLK_DIV1;
    RCC_ClkInitStruct.APB1CLKDivider = RCC_HCLK_DIV2;
    RCC_ClkInitStruct.APB2CLKDivider = RCC_HCLK_DIV1;

    if (HAL_RCC_ClockConfig(&RCC_ClkInitStruct, FLASH_LATENCY_2) != HAL_OK)
    {
        _Error_Handler(__FILE__, __LINE__);
    }

    // Configure the Systick interrupt time
    HAL_SYSTICK_Config(HAL_RCC_GetHCLKFreq()/1000);

    // Configure the Systick
    HAL_SYSTICK_CLKSourceConfig(SYSTICK_CLKSOURCE_HCLK);

    /* SysTick_IRQn interrupt configuration */
    HAL_NVIC_SetPriority(SysTick_IRQn, 0, 0);
}
```

```
/* USART2 init function */
static void MX_USART2_UART_Init(void)
{
  huart2.Instance = USART2;
  huart2.Init.BaudRate = 38400;
  huart2.Init.StopBits = UART_STOPBITS_1;
  huart2.Init.Parity = UART_PARITY_NONE;
  huart2.Init.Mode = UART_MODE_TX_RX;
  huart2.Init.HwFlowCtl = UART_HWCONTROL_NONE;
  huart2.Init.OverSampling = UART_OVERSAMPLING_16;
  huart2.Init.OneBitSampling = UART_ONE_BIT_SAMPLE_DISABLE;
  huart2.AdvancedInit.AdvFeatureInit = UART_ADVFEATURE_NO_INIT;
  if (HAL_UART_Init(&huart2) != HAL_OK)
  {
    _Error_Handler(__FILE__, __LINE__);
  }

}

// Enable DMA controller clock
static void MX_DMA_Init(void)
{
  /* DMA controller clock enable */
  __HAL_RCC_DMA1_CLK_ENABLE();

  // NOTE - No using interrupts for this demonstration
  /* DMA interrupt init */
  /* DMA1_Channel7_IRQn interrupt configuration */
  HAL_NVIC_SetPriority(DMA1_Channel7_IRQn, 0, 0);
  HAL_NVIC_EnableIRQ(DMA1_Channel7_IRQn);
}

/** Configure pins as
        * Analog
        * Input
        * Output
        * EVENT_OUT
        * EXTI
*/
static void MX_GPIO_Init(void)
{

  GPIO_InitTypeDef GPIO_InitStruct;

  /* GPIO Ports Clock Enable */
  __HAL_RCC_GPIOC_CLK_ENABLE();
  __HAL_RCC_GPIOF_CLK_ENABLE();
  __HAL_RCC_GPIOA_CLK_ENABLE();
  __HAL_RCC_GPIOB_CLK_ENABLE();
```

```
  /* Configure GPIO pin Output Level */
  HAL_GPIO_WritePin(LD2_GPIO_Port, LD2_Pin, GPIO_PIN_RESET);

  /*Configure GPIO pin : B1_Pin */
  GPIO_InitStruct.Pin = B1_Pin;
  GPIO_InitStruct.Mode = GPIO_MODE_IT_FALLING;
  GPIO_InitStruct.Pull = GPIO_NOPULL;
  HAL_GPIO_Init(B1_GPIO_Port, &GPIO_InitStruct);

  /*Configure GPIO pin : LD2_Pin */
  GPIO_InitStruct.Pin = LD2_Pin;
  GPIO_InitStruct.Mode = GPIO_MODE_OUTPUT_PP;
  GPIO_InitStruct.Pull = GPIO_NOPULL;
  GPIO_InitStruct.Speed = GPIO_SPEED_FREQ_LOW;
  HAL_GPIO_Init(LD2_GPIO_Port, &GPIO_InitStruct);
}

/* USER CODE BEGIN 4 */

/* USER CODE END 4 */

/**
  * @brief  This function is executed in case of error occurrence.
  * @param  None
  * @retval None
  */
void _Error_Handler(char * file, int line)
{
  /* USER CODE BEGIN Error_Handler_Debug */
  /* User can add his own implementation to report the HAL error return
state */
  while(1)
  {
  }
  /* USER CODE END Error_Handler_Debug */
}

#ifdef USE_FULL_ASSERT

/**
  * @brief Reports the name of the source file and the source line number
  * where the assert_param error has occurred.
  * @param file: pointer to the source file name
  * @param line: assert_param error line source number
  * @retval None
  */
void assert_failed(uint8_t* file, uint32_t line)
{
  /* USER CODE BEGIN 6 */
  /* User can add his own implementation to report the file name and
line number,
```

```
    ex: printf("Wrong parameters value: file %s on line %d\r\n", file,
line) */
  /* USER CODE END 6 */

}

#endif

/**
  * @}
  */

/**
  * @}
  */

/**** (C) COPYRIGHT STMicroelectronics ****/
```

I have first configured and initialized all the appropriate DMA setting in the main method. Some `DMA_InitTypeDef` struct members were not set because they are not needed in this demonstration and their values are not read during this particular DMA operation. The first three steps in the DMA preoperation sequence are as follows:

1. Set up appropriate memory and/or peripheral port addresses.

2. Establish the size of the data transfer.

3. Arm the DMA controller.

These steps were all accomplished using the following statement:

```
HAL_DMA_Start(&hdma_usart2_tx, (uint32_t)msg,
(uint32_t)&huart2.Instance->TDR, (uint32_t)strlen(msg));
```

The fourth and the final step is to start the DMA operation for the selected peripheral. This was accomplished by the following statement:

```
huart2.Instance->CR3 |= USART_CR3_DMAT;
```

The seventh bit in the USART2 control register CR3 is named DMAT and is set to commence normal DMA operations for this peripheral.

The core processor has also been set up to poll to determine when the DMA operation has been completed. This operation was accomplished by the following statement:

```
HAL_DMA_PollForTransfer(&hdma_usart2_tx, HAL_DMA_FULL_
TRANSFER, HAL_MAX_DELAY);
```

Polling for completion is completely appropriate for this demonstration simply because it is a "one-shot" occurrence and does not impede, nor conflict with any other MCU operations.

Finally, the USART2 peripheral is disarmed from further DMA operations by using the following statement:

```
huart2.Instance->CR3 &= ~USART_CR3_DMAT;
```

Test Results

I observed that the user LED did light after the hex file was downloaded into the project board. However, I desired a better indication that the DMA operation did run successfully. To do this, I started the Realterm application and monitored the virtual communications port that is logically connected to the USART2 peripheral. Figure 11-5 shows the screen capture displaying the data message sent out from the peripheral.

This absolutely confirmed that the USART2 peripheral transmitted the data message to memory via DMA.

This demonstration should now have convinced you of the DMA usefulness. It is time to introduce the DAC peripheral and really demonstrate how DMA can significantly improve MCU operations.

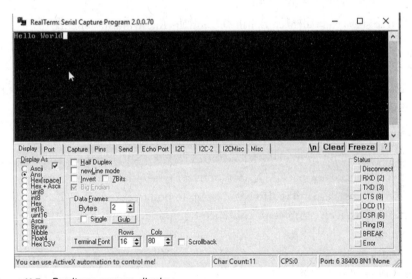

Figure 11-5   *Realterm screen display.*

DAC Peripheral

I will first explain the principles behind the DAC peripherals in order for you to understand how it requires rapid data transfers to operate efficiently.

DAC Principles

The DAC design used in STM MCUs is based on a R-2R resistor ladder network concept. This concept is easily explained by referring to Figure 11-6.

The resistive network acts like a programmable voltage divider between a reference voltage and ground. The usual reference voltage (V_{ref}) for the STM DAC is the main 3.3-V supply. There is a 3-bit DAC, shown Figure 11-6, which makes it easy to discuss. In reality the DAC used in the project board has a 12-bit input.

The binary inputs shown in Figure 11-6 are switched between 0 V and V_{ref} or 3.3 V. If a bit is high, it will cause its voltage input to weighted in the summed voltage output according to its bit position.

The output voltage (V_{out}) for the 3 bit ($N = 3$) example for a given digit binary input (Value) is determined by the following equation:

$$V_{out} = \frac{V_{ref} * Value}{2^N}$$

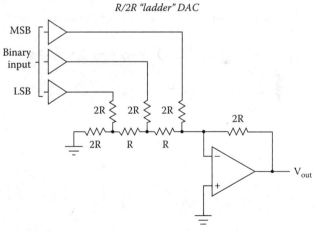

Figure 11-6   *R-2R design concept.*

Therefore, if the maximum value equals 7 or in binary terms 111, the maximum V_{out} will be:

$$V_{out} = \frac{3.3 * 7}{8} = 2.8875 \text{ V}$$

This result happens because the step size is 3.3/8 or 0.4125. This is a rather large error and only happens in this case due to the small number of bits involved. For the real 12-bit STM DAC, the step size would only be 3.3/4096 or 0.00080566 V or roughly 0.8 mV. That step size will provide an excellent analog voltage representation for a digital number. Returning to the digital example, I have calculated all eight analog voltage outputs for all the possible binary combinations. These are shown in Table 11-1.

Figure 11-7 shows the analog voltage output in more start terms where it becomes very apparent that there is no smooth transition between binary input values due to the limited number of bits.

Table 11-1 and Figure 11-7 provide more than ample proof that a sufficient number of bits must be present to have a somewhat smooth and continuous voltage output from a DAC. However, there will still be a small noise present even with a high number of bits present due to the inherent switching that is constantly happening with the DAC. This noise may be easily filtered out using an active low-pass analog filter. However, I will demonstrate that the DAC output is very clean and can be directly used as is without additional filtering.

A STM DAC also has the option of switching in a buffer output amplifier. This device will lower the DAC's output impedance and provide additional drive current without the need to add an external output operational amplifier for this purpose.

Digital Input—Value	Analog Output Voltage (V_{out})
000	0.0
001	0.4125
010	0.8250
011	1.2375
100	1.6500
101	2.0625
110	2.4750
111	2.8875

Table 11-1 *Binary Input vs Analog Voltage Output—3-bit DAC Example*

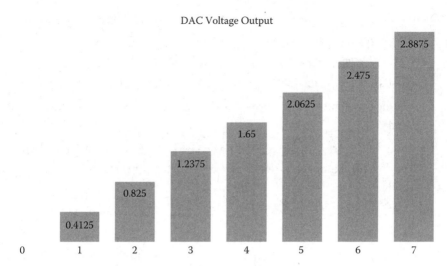

Figure 11-7 *Bar chart showing analog voltage outputs.*

It is now time to discuss the HAL software that can configure and initialize the STM DAC peripheral.

HAL Software for the DAC

The HAL framework uses a C struct named `DAC_HandleTypeDef` to configure the DAC peripheral. Each member will separately be discussed after the struct definition.

```
typedef struct {
DAC_TypeDef                  *Instance;
__IO HAL_DAC_StateTypeDef    State;
HAL_LockTypeDef              Lock;
DMA_HandleTypeDef            *DMA_Handle1;
DMA_HandleTypeDef            *DMA_Handle2;
__IO uint32_t                ErrorCode;
} DAC_HandleTypeDef;
```

The `DAC_HandleTypeDef` struct members are briefly defined in the following list:

- `DAC_TypeDef *Instance;`—This is a pointer to the DAC descriptor which will be used. For example, DAC1 is the descriptor for the first DAC peripheral.

- `__IO HAL_DAC_StateTypeDef State;`—This value represents the DAC communication state.

- `HAL_LockTypeDef Lock;`—This value is a DAC-locking object.

- `DMA_HandleTypeDef *DMA_Handle1;`—This is a pointer to a DMA handler configured to perform digital-to-analog conversions in DMA mode.

- `DMA_HandleTypeDef *DMA_Handle2;`—This is a pointer to a DMA handler configured to perform digital-to-analog conversions in DMA mode.

- `__IO uint32_t ErrorCode;`—This is a value representing the DAC error code.

The `DAC_HandleTypeDef` struct differs from all the other similar structs in that it does not have an Init member, but instead relies on a different type of C struct to initialize the DAC at the channel level. This C sruct is named `DAC_ChannelConfTypeDef` and has the following definition. As before, each member will separately be discussed after the struct definition.

```
typedef struct {
uint32_t    DAC_Trigger;
uint32_t    DAC_OutputBuffer;
} DAC_ChannelConfTypeDef;
```

The `DAC_ChannelConfTypeDef` struct members are briefly defined in the following list:

- `uint32_t DAC_Trigger;`—This value specifies the source used to trigger the DAC conversion. This parameter can be any value of the following defines:

  - `DAC_TRIGGER_NONE`—Used for a manual trigger.

  - `DAC_TRIGGER_SOFTWARE`—Used when DAC is in a DMA mode without a timer.

  - `DAC_TRIGGER_Tx_TRGO`—Used when DAC is driven by a dedicated timer.

- `uint32_t DAC_OutputBuffer;`—Specifies whether the DAC channel output buffer is enabled or disabled.

Demonstration Two

In this demonstration I will be manually triggering the DAC to output a repetitive voltage ramp.

Program Listing

The following is the code listing for this demonstration. I have also provided some additional discussions regarding pertinent code segments following the listing.

```
// STM disclaimer goes here
// Includes
#include "main.h"
#include "stm32f3xx_hal.h"

/* USER CODE BEGIN Includes */

/* USER CODE END Includes */

// Private variables
DAC_HandleTypeDef hdac;

/* USER CODE BEGIN PV */
// Private variables

/* USER CODE END PV */

// Private function prototypes
void SystemClock_Config(void);
static void MX_GPIO_Init(void);
static void MX_DAC_Init(void);

/* USER CODE BEGIN PFP */
// Private function prototypes

/* USER CODE END PFP */

/* USER CODE BEGIN 0 */

/* USER CODE END 0 */

int main(void)
{

  /* USER CODE BEGIN 1 */

  /* USER CODE END 1 */
```

```
// MCU Configuration

// Reset all peripherals, initializes the flash and Systick.
HAL_Init();

/* USER CODE BEGIN Init */

/* USER CODE END Init */

/* Configure the system clock */
SystemClock_Config();

/* USER CODE BEGIN SysInit */

/* USER CODE END SysInit */

/* Initialize all configured peripherals */
MX_GPIO_Init();
MX_DAC_Init();

/* USER CODE BEGIN 2 */
HAL_DAC_Init(&hdac);
HAL_DAC_Start(&hdac, DAC_CHANNEL_1);
/* USER CODE END 2 */

/* Infinite loop */
/* USER CODE BEGIN WHILE */
while (1)
{
/* USER CODE END WHILE */

/* USER CODE BEGIN 3 */
    int i =2000;
        while(i <4000) {
           HAL_DAC_SetValue(&hdac, DAC_CHANNEL_1, DAC_ALIGN_12B_R, i);
           HAL_Delay(0.001);
           i +=4;
        }
        while(i > 2000) {
           HAL_DAC_SetValue(&hdac, DAC_CHANNEL_1, DAC_ALIGN_12B_R, i);
           HAL_Delay(0.001);
           i -=4;
        }
}
    /* USER CODE END 3 */

}

// System Clock Configuration
void SystemClock_Config(void)
{
```

```
    RCC_OscInitTypeDef RCC_OscInitStruct;
    RCC_ClkInitTypeDef RCC_ClkInitStruct;

    // Initializes the CPU, AHB and APB buss clocks
    RCC_OscInitStruct.OscillatorType = RCC_OSCILLATORTYPE_HSI;
    RCC_OscInitStruct.HSIState = RCC_HSI_ON;
    RCC_OscInitStruct.HSICalibrationValue = 16;
    RCC_OscInitStruct.PLL.PLLState = RCC_PLL_ON;
    RCC_OscInitStruct.PLL.PLLSource = RCC_PLLSOURCE_HSI;
    RCC_OscInitStruct.PLL.PLLMUL = RCC_PLL_MUL16;
    if (HAL_RCC_OscConfig(&RCC_OscInitStruct) != HAL_OK)
    {
      _Error_Handler(__FILE__, __LINE__);
    }

    // Initialize the CPU, AHB and APB buss clocks
    RCC_ClkInitStruct.ClockType =
    RCC_CLOCKTYPE_HCLK|RCC_CLOCKTYPE_SYSCLK
    |RCC_CLOCKTYPE_PCLK1|RCC_CLOCKTYPE_PCLK2;
    RCC_ClkInitStruct.SYSCLKSource = RCC_SYSCLKSOURCE_PLLCLK;
    RCC_ClkInitStruct.AHBCLKDivider = RCC_SYSCLK_DIV1;
    RCC_ClkInitStruct.APB1CLKDivider = RCC_HCLK_DIV2;
    RCC_ClkInitStruct.APB2CLKDivider = RCC_HCLK_DIV1;

    if (HAL_RCC_ClockConfig(&RCC_ClkInitStruct, FLASH_LATENCY_2) != HAL_OK)
    {
      _Error_Handler(__FILE__, __LINE__);
    }

    // Configure the Systick interrupt time
    HAL_SYSTICK_Config(HAL_RCC_GetHCLKFreq()/1000);

    // Configure the Systick
    HAL_SYSTICK_CLKSourceConfig(SYSTICK_CLKSOURCE_HCLK);

    /* SysTick_IRQn interrupt configuration */
    HAL_NVIC_SetPriority(SysTick_IRQn, 0, 0);
}
/* DAC init function */
static void MX_DAC_Init(void)
{

    DAC_ChannelConfTypeDef sConfig;

    // DAC Initialization
    hdac.Instance = DAC;
    if (HAL_DAC_Init(&hdac) != HAL_OK)
    {
      _Error_Handler(__FILE__, __LINE__);
    }
```

```c
  // DAC channel OUT1 config
  sConfig.DAC_Trigger = DAC_TRIGGER_NONE;
  sConfig.DAC_OutputBuffer = DAC_OUTPUTBUFFER_ENABLE;
  if (HAL_DAC_ConfigChannel(&hdac, &sConfig, DAC_CHANNEL_1) != HAL_OK)
  {
    _Error_Handler(__FILE__, __LINE__);
  }
}

/** Configure pins as
        * Analog
        * Input
        * Output
        * EVENT_OUT
        * EXTI
     PA2    ------> USART2_TX
     PA3    ------> USART2_RX
*/
static void MX_GPIO_Init(void)
{
  GPIO_InitTypeDef GPIO_InitStruct;

  /* GPIO Ports Clock Enable */
  __HAL_RCC_GPIOC_CLK_ENABLE();
  __HAL_RCC_GPIOF_CLK_ENABLE();
  __HAL_RCC_GPIOA_CLK_ENABLE();
  __HAL_RCC_GPIOB_CLK_ENABLE();

  /*Configure GPIO pin Output Level */
  HAL_GPIO_WritePin(LD2_GPIO_Port, LD2_Pin, GPIO_PIN_RESET);

  /*Configure GPIO pin : B1_Pin */
  GPIO_InitStruct.Pin = B1_Pin;
  GPIO_InitStruct.Mode = GPIO_MODE_IT_FALLING;
  GPIO_InitStruct.Pull = GPIO_NOPULL;
  HAL_GPIO_Init(B1_GPIO_Port, &GPIO_InitStruct);

  /*Configure GPIO pins : USART_TX_Pin USART_RX_Pin */
  GPIO_InitStruct.Pin = USART_TX_Pin|USART_RX_Pin;
  GPIO_InitStruct.Mode = GPIO_MODE_AF_PP;
  GPIO_InitStruct.Pull = GPIO_NOPULL;
  GPIO_InitStruct.Speed = GPIO_SPEED_FREQ_LOW;
  GPIO_InitStruct.Alternate = GPIO_AF7_USART2;
  HAL_GPIO_Init(GPIOA, &GPIO_InitStruct);

  /*Configure GPIO pin : LD2_Pin */
  GPIO_InitStruct.Pin = LD2_Pin;
  GPIO_InitStruct.Mode = GPIO_MODE_OUTPUT_PP;
  GPIO_InitStruct.Pull = GPIO_NOPULL;
```

```
    GPIO_InitStruct.Speed = GPIO_SPEED_FREQ_LOW;
    HAL_GPIO_Init(LD2_GPIO_Port, &GPIO_InitStruct);
}

/* USER CODE BEGIN 4 */

/* USER CODE END 4 */

/**
  * @brief  This function is executed in case of error occurrence.
  * @param  None
  * @retval None
  */
void _Error_Handler(char * file, int line)
{
  /* USER CODE BEGIN Error_Handler_Debug */
  /* User can add his own implementation to report the HAL error return
state */
  while(1)
  {
  }
  /* USER CODE END Error_Handler_Debug */
}

#ifdef USE_FULL_ASSERT

/**
  * @brief Reports the name of the source file and the source line number
  * where the assert_param error has occurred.
  * @param file: pointer to the source file name
  * @param line: assert_param error line source number
  * @retval None
  */
void assert_failed(uint8_t* file, uint32_t line)
{
  /* USER CODE BEGIN 6 */
  /* User can add his own implementation to report the file name and
line number,
     ex: printf("Wrong parameters value: file %s on line %d\r\n", file,
line) */
  /* USER CODE END 6 */

}

#endif

/**
  * @}
  */
```

```
/**
  * @}
  */
```

```
/**** (C) COPYRIGHT STMicroelectronics ****/
```

The significant addition to this basic code is the DAC Init function MX_DAC_ Init. The CubeMX application generated this code when the DAC peripheral was enabled during the project creation process. There is really nothing remarkable that is happening in this function other than the output buffer is enabled and any DAC trigger is disabled.

The functional code that produces an output waveform is located in the forever loop of the main method. This code segment is relisted below to allow for an easier reference:

```
while (1)
  {
  /* USER CODE END WHILE */
  /* USER CODE BEGIN 3 */
    int i =2000;
        while(i <4000) {
        HAL_DAC_SetValue(&hdac, DAC_CHANNEL_1, DAC_ALIGN_12B_R, i);
        HAL_Delay(0.001);
        i +=4;
        }
        while(i > 2000) {
        HAL_DAC_SetValue(&hdac, DAC_CHANNEL_1, DAC_ALIGN_12B_R, i);
        HAL_Delay(0.001);
        i -=4;
        }
        }
  /* USER CODE END 3 */
}
```

There are two while loops within the forever loop, one which counts up and outputs a slowly increasing voltage and other counts down and outputs a slowly decreasing voltage. The waveform period is determined by the delay value, which in this demonstration has been set to 1 ms between each digital incremental value. The HAL_DAC_SetValue method causes the DAC to output on DAC channel 1 an analog voltage based on the 12-bit, right endian justified value for the index i.

Test Results

Figure 11-8 is a screen shot from a USB oscilloscope connected to PA4, which is the DAC channel 1 output for the STM32F302R8 MCU.

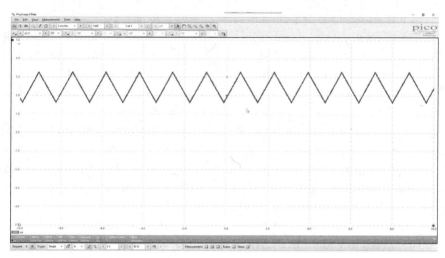

Figure 11-8   *DAC output.*

You can see that it is relatively clean triangular waveform with a 1.626-ms period or an approximate frequency equal to 615 Hz. The peak-to-peak amplitude varies between approximately 1.6 and 3.3 V. These particular voltages are a result of using the values 2000 and 4000 as the DAC input limits, respectively.

I am reasonably sure that you can also increase the waveform frequency by decreasing the delay in the code as described above. Readers should try experimenting to determine the ultimate frequency limit.

Demonstration Three

In this last demonstration I will be operating the DAC in a DMA mode to output a sine wave that is synthesized by the application. Generating an analog signal is a common task assigned to a DAC. DMA is the preferred operational method because it frees up the core processor to attend to higher priority tasks rather than focusing on generating binary input for the DAC. The following generalized statement both starts the DAC and performs a DMA data transfer from memory to the peripheral:

```
HAL_StatusTypeDef HAL_DAC_Start_DMA(DAC_HandleTypeDef*
hdac, uint32_t Channel, uint32_t* pData, uint32_t
Length, uint32_t Alignment);
```

In this demonstration, only one DAC channel named DAC_CHANNEL_1 is available. The argument uint32_t* pData is a data pointer, which in this case will point to the beginning of the 200 element SINE array. The uint32_t Length argument is the number of data elements to be transferred, which is the NSAMP constant with value 200. Finally, the uint32_t Alignment argument specifies the data element bit size and whether the data is right or left justified. In this case, the define DAC_ALIGN_12B_R shows the data is 12-bit and is aligned to the right, meaning that the least significant data bit is also the rightmost bit.

Synthesizing a sine wave requires setting the number of increments or steps that will go into making up a single cycle waveform. In this example, the number was set at 200, which is a good trade-off between having a representative sample and minimizing the total steps that need to be constantly handled. The resulting equation used to synthesize a 12-bit sine wave is as follows:

$$DAC_{input} = \left(\sin\left(i \times \frac{2\pi}{NSAMP} \right) + 1 \right) \times \left(\frac{4096}{2} \right)$$

This equation was used to synthesize a single sine wave cycle with the i step ranging from 0 to 199.

I next arbitrarily selected a period of 15 ms for the DAC sine wave output or an equivalent frequency of 66.67 Hz. This means that the conversion rate must be set at 66.67 * 200 or 13,334.

The timer TIM6 is used as trigger for the DAC's TRGO line and its period must then be set to 13,334. The high-speed external clock (HSE) input into TIM6 was set at 64 MHz with no prescaling. This all means the timer period value must be set to 64 MHz/13.33 KHz or approximately 4800. The actual period value was set at 4799 to account for the added 1 that goes into to the timer frequency calculation.

Program Listing

The following is the code listing for this demonstration. I have also provided some additional discussions regarding pertinent code segments following the listing.

```
// The STM disclaimer goes here
// Includes
#include "main.h"
#include "stm32f3xx_hal.h"
#include <math.h>

/* USER CODE BEGIN Includes */
```

```
/* USER CODE END Includes */

#define PI   3.14159265
#define NSAMP  200

// Private variables
DAC_HandleTypeDef hdac;
TIM_HandleTypeDef htim6;
DMA_HandleTypeDef hdma_dac_ch1;

/* USER CODE BEGIN PV */
// Private variables

/* USER CODE END PV */

// Private function prototypes
void SystemClock_Config(void);
static void MX_GPIO_Init(void);
static void MX_DMA_Init(void);
static void MX_DAC_Init(void);
static void MX_TIM6_Init(void);

/* USER CODE BEGIN PFP */
// Private function prototypes

/* USER CODE END PFP */

/* USER CODE BEGIN 0 */

/* USER CODE END 0 */

int main(void)
{

  /* USER CODE BEGIN 1 */
  uint16_t SINE[NSAMP], value;
  /* USER CODE END 1 */

  // MCU Configuration

  // Reset all peripherals, initializes the flash and Systick.
  HAL_Init();

  /* USER CODE BEGIN Init */

  /* USER CODE END Init */

  /* Configure the system clock */
  SystemClock_Config();
```

```
    /* USER CODE BEGIN SysInit */

    /* USER CODE END SysInit */

    /* Initialize all configured peripherals */
    MX_GPIO_Init();
    MX_DMA_Init();
    MX_DAC_Init();
    MX_TIM6_Init();

    /* USER CODE BEGIN 2 */
    for (uint16_t i = 0; i < NSAMP; i++) {
        value = (uint16_t) rint((sinf(((2*PI)/NSAMP)*i)+1)*2048);
            SINE[i] = value < 4096 ? value : 4095;
    }
    /* USER CODE END 2 */

    HAL_DAC_Init(&hdac);
    HAL_TIM_Base_Start(&htim6);
    HAL_DAC_Start_DMA(&hdac, DAC_CHANNEL_1, (uint32_t*)SINE, NSAMP, DAC_
ALIGN_12B_R);

/* Infinite loop */
    /* USER CODE BEGIN WHILE */
    while (1)
    {
    /* USER CODE END WHILE */

    /* USER CODE BEGIN 3 */

    }
    /* USER CODE END 3 */
}

// System Clock Configuration
void SystemClock_Config(void)
{
    RCC_OscInitTypeDef RCC_OscInitStruct;
    RCC_ClkInitTypeDef RCC_ClkInitStruct;

    // Initialize the CPU, AHB and APB buss clocks
    RCC_OscInitStruct.OscillatorType = RCC_OSCILLATORTYPE_HSI;
    RCC_OscInitStruct.HSIState = RCC_HSI_ON;
    RCC_OscInitStruct.HSICalibrationValue = 16;
    RCC_OscInitStruct.PLL.PLLState = RCC_PLL_ON;
    RCC_OscInitStruct.PLL.PLLSource = RCC_PLLSOURCE_HSI;
    RCC_OscInitStruct.PLL.PLLMUL = RCC_PLL_MUL16;
    if (HAL_RCC_OscConfig(&RCC_OscInitStruct) != HAL_OK)
    {
        _Error_Handler(__FILE__, __LINE__);
    }
```

```
  // Initialize the CPU, AHB and APB buss clocks
  RCC_ClkInitStruct.ClockType =
  RCC_CLOCKTYPE_HCLK|RCC_CLOCKTYPE_SYSCLK
  |RCC_CLOCKTYPE_PCLK1|RCC_CLOCKTYPE_PCLK2;
  RCC_ClkInitStruct.SYSCLKSource = RCC_SYSCLKSOURCE_PLLCLK;
  RCC_ClkInitStruct.AHBCLKDivider = RCC_SYSCLK_DIV1;
  RCC_ClkInitStruct.APB1CLKDivider = RCC_HCLK_DIV2;
  RCC_ClkInitStruct.APB2CLKDivider = RCC_HCLK_DIV1;

  if (HAL_RCC_ClockConfig(&RCC_ClkInitStruct, FLASH_LATENCY_2) !=
HAL_OK)
  {
    _Error_Handler(__FILE__, __LINE__);
  }

  // Configure the Systick interrupt time
  HAL_SYSTICK_Config(HAL_RCC_GetHCLKFreq()/1000);

  // Configure the Systick
  HAL_SYSTICK_CLKSourceConfig(SYSTICK_CLKSOURCE_HCLK);

  /* SysTick_IRQn interrupt configuration */
  HAL_NVIC_SetPriority(SysTick_IRQn, 0, 0);
}

/* DAC init function */
static void MX_DAC_Init(void)
{
  DAC_ChannelConfTypeDef sConfig;
  GPIO_InitTypeDef GPIO_InitStruct;

  __HAL_RCC_DAC1_CLK_ENABLE();

  // DAC Initialization
  hdac.Instance = DAC;
  if (HAL_DAC_Init(&hdac) != HAL_OK)
  {
    _Error_Handler(__FILE__, __LINE__);
  }

  // DAC channel OUT1 config
  sConfig.DAC_Trigger = DAC_TRIGGER_T6_TRGO;
  sConfig.DAC_OutputBuffer = DAC_OUTPUTBUFFER_ENABLE;
  if (HAL_DAC_ConfigChannel(&hdac, &sConfig, DAC_CHANNEL_1) != HAL_OK)
  {
    _Error_Handler(__FILE__, __LINE__);
  }

  /** DAC GPIO configuration */
  GPIO_InitStruct.Pin = GPIO_PIN_4;
```

```
  GPIO_InitStruct.Mode = GPIO_MODE_ANALOG;
  GPIO_InitStruct.Pull = GPIO_NOPULL;
  HAL_GPIO_Init(GPIOA, &GPIO_InitStruct);

  /** Peripheral DMA init */
  hdma_dac_ch1.Instance = DMA1_Channel3;
  hdma_dac_ch1.Init.Direction = DMA_MEMORY_TO_PERIPH;
  hdma_dac_ch1.Init.PeriphInc = DMA_PINC_DISABLE;
  hdma_dac_ch1.Init.MemInc = DMA_MINC_ENABLE;
  hdma_dac_ch1.Init.PeriphDataAlignment = DMA_PDATAALIGN_HALFWORD;
  hdma_dac_ch1.Init.MemDataAlignment = DMA_MDATAALIGN_HALFWORD;
  hdma_dac_ch1.Init.Mode = DMA_CIRCULAR;
  hdma_dac_ch1.Init.Priority = DMA_PRIORITY_LOW;
  HAL_DMA_Init(&hdma_dac_ch1);

  __HAL_LINKDMA(&hdac, DMA_Handle1, hdma_dac_ch1);
}

/* TIM6 init function */
static void MX_TIM6_Init(void)
{
  TIM_MasterConfigTypeDef sMasterConfig;

  __HAL_RCC_TIM6_CLK_ENABLE();

  htim6.Instance = TIM6;
  htim6.Init.Prescaler = 0;
  htim6.Init.CounterMode = TIM_COUNTERMODE_UP;
  htim6.Init.Period = 4799;
  htim6.Init.AutoReloadPreload = TIM_AUTORELOAD_PRELOAD_DISABLE;
  if (HAL_TIM_Base_Init(&htim6) != HAL_OK)
  {
    _Error_Handler(__FILE__, __LINE__);
  }

  sMasterConfig.MasterOutputTrigger = TIM_TRGO_UPDATE;
  sMasterConfig.MasterSlaveMode = TIM_MASTERSLAVEMODE_DISABLE;
  if (HAL_TIMEx_MasterConfigSynchronization(&htim6, &sMasterConfig) !=
HAL_OK)
  {
    _Error_Handler(__FILE__, __LINE__);
  }
}

// Enable DMA controller clock
static void MX_DMA_Init(void)
{
  /* DMA controller clock enable */
  __HAL_RCC_DMA1_CLK_ENABLE();

  /* DMA interrupt init */
```

```
  /* DMA1_Channel3_IRQn interrupt configuration */
  HAL_NVIC_SetPriority(DMA1_Channel3_IRQn, 0, 0);
  HAL_NVIC_EnableIRQ(DMA1_Channel3_IRQn);
}

/** Configure pins as
        * Analog
        * Input
        * Output
        * EVENT_OUT
        * EXTI
     PA2     ------> USART2_TX
     PA3     ------> USART2_RX
*/

static void MX_GPIO_Init(void)
{
  GPIO_InitTypeDef GPIO_InitStruct;

  /* GPIO Ports Clock Enable */
  __HAL_RCC_GPIOC_CLK_ENABLE();
  __HAL_RCC_GPIOF_CLK_ENABLE();
  __HAL_RCC_GPIOA_CLK_ENABLE();
  __HAL_RCC_GPIOB_CLK_ENABLE();

  /*Configure GPIO pin Output Level */
  HAL_GPIO_WritePin(LD2_GPIO_Port, LD2_Pin, GPIO_PIN_RESET);

  /*Configure GPIO pin : B1_Pin */
  GPIO_InitStruct.Pin = B1_Pin;
  GPIO_InitStruct.Mode = GPIO_MODE_IT_FALLING;
  GPIO_InitStruct.Pull = GPIO_NOPULL;
  HAL_GPIO_Init(B1_GPIO_Port, &GPIO_InitStruct);

  /*Configure GPIO pins : USART_TX_Pin USART_RX_Pin */
  GPIO_InitStruct.Pin = USART_TX_Pin|USART_RX_Pin;
  GPIO_InitStruct.Mode = GPIO_MODE_AF_PP;
  GPIO_InitStruct.Pull = GPIO_NOPULL;
  GPIO_InitStruct.Speed = GPIO_SPEED_FREQ_LOW;
  GPIO_InitStruct.Alternate = GPIO_AF7_USART2;
  HAL_GPIO_Init(GPIOA, &GPIO_InitStruct);

  /*Configure GPIO pin : LD2_Pin */
  GPIO_InitStruct.Pin = LD2_Pin;
  GPIO_InitStruct.Mode = GPIO_MODE_OUTPUT_PP;
  GPIO_InitStruct.Pull = GPIO_NOPULL;
  GPIO_InitStruct.Speed = GPIO_SPEED_FREQ_LOW;
  HAL_GPIO_Init(LD2_GPIO_Port, &GPIO_InitStruct);

}

/* USER CODE BEGIN 4 */
```

```
/* USER CODE END 4 */

/**
  * @brief  This function is executed in case of error occurrence.
  * @param  None
  * @retval None
  */
void _Error_Handler(char * file, int line)
{
  /* USER CODE BEGIN Error_Handler_Debug */
  /* User can add his own implementation to report the HAL error re-
turn state */
  while(1)
  {
  }
  /* USER CODE END Error_Handler_Debug */
}

#ifdef USE_FULL_ASSERT

/**
   * @brief Reports the name of the source file and the source line
number
   * where the assert_param error has occurred.
   * @param file: pointer to the source file name
   * @param line: assert_param error line source number
   * @retval None
   */
void assert_failed(uint8_t* file, uint32_t line)
{
  /* USER CODE BEGIN 6 */
  /* User can add his own implementation to report the file name and
line number,
     ex: printf("Wrong parameters value: file %s on line %d\r\n", file,
line) */
  /* USER CODE END 6 */

}

#endif

/**
  * @}
  */

/**
   * @}
*/

/**** (C) COPYRIGHT STMicroelectronics ****/
```

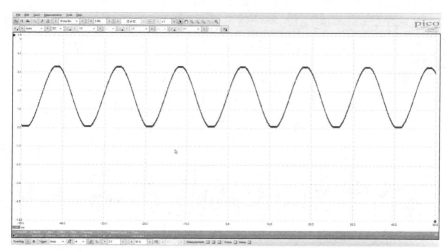

Figure 11-9   *DAC output.*

Test Results

Figure 11-9 is a screen shot from a USB oscilloscope connected to PA4, which is the DAC channel 1 output for the STM32F302R8 MCU.

You can see that it is a relatively clean sine waveform with a 15-ms period or an approximate frequency equal to 66.67 Hz. The peak-to-peak amplitude varies between approximately 0 and 3.3 V. This waveform is exactly what was programmed.

Figure 11-10 is an averaged spectrum display, which shows the main frequency components in the DAC sine wave output signal.

The average total harmonic distortion (THD) was measured at only 1.372%, which is very low for this type of signal generation. The third harmonic at 200 Hz was the highest distortion component and was measured at −42 dBu, which is less than one-hundredth of the primary frequency's amplitude.

I also experimented by changing the timer TIM6 period to determine the highest possible output frequency. I determined that a minimum period value of 105 created a sine wave frequency slightly over 3 kHz. Any value lower than that period value simply caused the DAC to remain at steady high level.

The last point I will make, which is the same one that I have often repeated, is that the core Cortex-M processor is not involved with this signal generation once it has been configured, initialized, and started. Notice that the forever loop in the

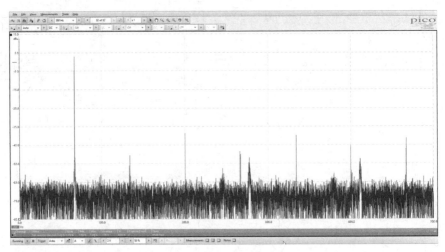

Figure 11-10 *Sine wave spectrum display.*

main method is completely devoid of any functional code because the signal generation is completely being accomplished by a combination of hardware and firmware. This approach is the essence of efficient and effective embedded design.

Summary

This is a combination chapter that explains both direct memory access (DMA)
and the digital-to-analog converter (DAC). I chose this combination because the
DAC is an ideal peripheral to use with DMA operations.

The first section explains the theory behind DMA and how a DMA controller
effectively implements this operational mode.

The next discussion focuses on how to use the HAL framework to both configure and initialize the MCU for DMA operations. I presented a detailed discussion
concerning two C structs that are critical for setting up the DMA mode.

The first demonstration simply has the USART2 peripheral sending a character string to memory using the DMA mode. A complete program listing was provided with all the important details concerning DMA operations highlighted.

The next section introduces the DAC and provides a brief background on how
it functions. This is followed by a section on how to use the HAL framework to
configure and initialize this peripheral.

The second demonstration illustrates how to have the DAC generate a triangular waveform using functional code placed in the main method's forever loop. This demonstration was purposefully designed to be computationally "heavy" in anticipation of the next demonstration.

The third and final demonstration shows how the DAC could generate a relatively "pure" sine wave using a DMA mode and then being periodically triggered by a timer. The initial signal was set at a 66-Hz frequency, but I eventually was able to generate sine wave signals slightly exceeding 3 kHz. The signal waveforms were generated without any core processor involvement other than the initial configuration.

INDEX

/ /, 96
/* ... */, 96
#, 96
{ }, 96, 97

0-ohm resistors, 8
0x, 103
0x0, 148
0xffff, 148
0xffff ffff, 148
3.3-V external power supply input, 26, 26t
10x high impedance probe settings, 127
19.4-MHz bandwidth, 127
32-bit STM address range, 103, 104f
50% duty cycle waveform, 227–228, 229t
70-MHz bandwidth, 127

acknowledgment instruction, 133
ADC. *See* analog-to-digital converter (ADC)
ADC clock speed, 194
ADC conversion modes, 197–198
ADC resolution, 194
ADC_CHANNEL_0 to ADC_CHANNEL_15, 199–200
ADC_ChannelConfTypeDef 199
ADC_CHANNEL_VBAT, 199–200
ADC_CHANNEL_VREFINT, 199–200
ADC_CLOCK_PCLK DIV1, 194
ADC_CLOCK_PCLK DIV2, 194
ADC_CLOCK_PCLK DIV4, 194
ADC_CLOCK_PCLK DIV6, 194
ADC_CLOCK_PCLK DIV8, 194
ADC_DATAALIGN_LEFT, 194
ADC_DATAALIGN_RIGHT, 194
ADC_EOC_SEQ_CONV, 195
ADC_EOC_SINGLE_CONV, 195
ADC_EXTERNALTRIGCONVEDGE_FALLING, 196
ADC_EXTERNALTRIGCONVEDGE_NONE, 196
ADC_EXTERNALTRIGCONVEDGE_RISING, 196
ADC_EXTERNALTRIGCONVEDGE_
    RISINGFALLING, 196
ADC_EXTERNALTRIGCONV_EXT_IT11, 196
ADC_EXTERNALTRIGCONV_T1_CC1, 196
ADC_EXTERNALTRIGCONV_T1_CC2, 196
ADC_EXTERNALTRIGCONV_T1_CC3, 196
ADC_EXTERNALTRIGCONV_T2_CC2, 196
ADC_EXTERNALTRIGCONV_T3_CC4, 196
ADC_EXTERNALTRIGCONV_T4_CC4, 196
ADC_EXTERNALTRIGCONV_T1_TRGO, 196
ADC_EXTERNALTRIGCONV_T1_TRGO2, 196
ADC_EXTERNALTRIGCONV_T2_TRGO, 196
ADC_EXTERNALTRIGCONV_T4_TRGO, 196

ADC_EXTERNALTRIGCONV_T5_TRGO, 196
ADC_EXTERNALTRIGCONV_T6_TRGO, 196
ADC_EXTERNALTRIGCONV_T8_TRGO, 196
ADC_EXTERNALTRIGCONV_T8_TRGO2, 196
ADC_HandleTypeDef, 192–193
ADC_InitTypeDef C struct, 193–194
ADC_InitTypeDef Init, 192, 193
ADC_REGULAR_RANK_1 to ADC_REGULAR_
    RANK_16, 199–200
ADC_RESOLUTION_6B, 194
ADC_RESOLUTION_12B, 194
ADC_SAMPLINGTIME_3CYCLES, 200
ADC_SAMPLINGTIME_15CYCLES, 200
ADC_SAMPLINGTIME_28CYCLES, 200
ADC_SAMPLINGTIME_56CYCLES, 200
ADC_SAMPLINGTIME_84CYCLES, 200
ADC_SAMPLINGTIME_112CYCLES, 200
ADC_SAMPLINGTIME_144CYCLES, 200
ADC_SAMPLINGTIME_480CYCLES, 200
ADC_SCAN_DISABLE, 194
ADC_SCAN_ENABLE, 194
ADC_SOFTWARE START, 196
ADC_TypeDef *Instance, 192, 193
advanced control timer (TIM1), 17, 17t, 146, 147t
AdvancedInit, 170
AHB, 244f
AHB1, 101, 106f
AHB2, 106f
AHB3, 106f
aliased memory, 105t
Analog Devices TMP36 temperature sensor, 189, 190f,
    202–205
analog servo, 235–236, 236f
analog-to-digital conversion, 189–212
    ADC, 16, 189
    ADC clock speed, 194
    ADC conversion modes, 197–198
    ADC functions, 190–192
    ADC resolution, 194
    channels, groups, and ranks, 198–200
    demonstration program, 200–212
    discontinuous/continuous mode, 195
    DMA requests, 196
    end of conversion (EOC), 195
    group, 198
    HAL framework, 192–197
    multi-channel scan/ continuous conversion, 198
    multi-channel scan/single conversion, 197–198, 198f
    polling, 200, 201
    rank, 199, 199f
    single channel/continuous conversion, 198

analog-to-digital conversion (*Cont.*)
  single channel/single conversion, 197, 197f
  successive-approximation register (SAR), 190, 191f
  TMP36 temperature sensor, 189, 190f, 202–205
analog-to-digital converter (ADC), 16, 189
analog voltage outputs, 260, 260t, 261f
analog voltage reference (AREF), 12
APB1 peripheral allocations, 107, 108f
APB2, 106f
Application/MDK-ARM, 70f, 80
Application/User, 70f, 80
Arduino connectors, 10–12, 114f
Arduino IDE, 32
Arduino prototype board, 112, 114f
AREF. *See* analog voltage reference (AREF)
ARM, 78
ARM Cortex M-4 block diagram, 5f
ARM Cortex Microcontroller Interface Standard
  (CMSIS), 78–85
  axf format, 82
  Build process log, 81, 81f
  compiling and downloading the project, 80–82
  components, 79
  CubeMX-generated C code, 80, 80f
  downloading the hex code, 82–86
  Keil IDE build icon, 81, 81f
  ST-LINK, 82–86
arming/disarming, 131
asynchronous event, 129
axf format, 82

B1 USER, 22
B2 RESET, 22
background thread/foreground thread, 132
Bare metal development, 30
BASEPRI register, 131–132
basic timer (TIM6), 17t, 18, 146, 147t
baud rate chart, 171f
Baudrate, 171
binary code, 34
binary divider (prescaler), 145, 147–148
bit serial communications, 167–188
  clear-to-send (CTS) line, 169
  data transfer protocol, 168–169, 188
  enabling USART2, 175–176
  full-duplex communications, 168, 169f
  protocol control lines, 168–169
  Realterm terminal program, 172–175
  request/acknowledge, 169
  request-to-send (RTS) line, 169
  UART block diagram, 167f
  UART vs. USART, 167, 168
  USART block diagram, 168f
  USART configuration, 170–172
  USART2 global interrupt, 182
  USART receive demonstration program,
    182–188
  USART transmit demonstration program,
    176–181
  Windows terminal program, 172–175

bit serial ports, 19–21
  I2C, 19–20
  I2S, 21
  SPI, 20–21
  USART, 19
Board Selector dialog screen shot, 55, 57f
BOOT, 58f, 59
BOOT0 input pin, 13
braces { }, 96, 97
buffer output amplifier, 260
bugs, 34
Build process log, 81, 81f
bus controller, 241, 244
bus matrix, 244, 244f, 245
bus matrix block diagram, 244f

C/C++ language projects. *See* STM project
  development
CCRx register. *See* compare capture registers
  (CCRx)
CIE1931 lightness formula, 231
clear-to-send (CTS) line, 169
Clock Configuration gui, 62f
Clock Configuration tab, 62–63
clock skew, 124–125, 128
clock speed, 124
clock speed demonstration, 124–127
ClockDivision, 148
CMSIS. *See* ARM Cortex Microcontroller Interface
  Standard (CMSIS)
CMSIS-CORE, 79
CMSIS-DAP, 79
CMSIS-Driver, 79
CMSIS-DSP, 79
CMSIS-Pack, 79
CMSIS-RTOS, 79
CMSIS-SVD, 79
CN5 connector pinout, 11, 12t
CN6 connector pinout, 10f, 11t
CN7 connector pinout, 13, 13t
CN8 connector pinout, 11, 11t
CN9 connector pinout, 12, 12t
CN10 connector pinout, 13, 13t
CNT counter, 216, 217f
Code Generation dialog box, 68, 69f
Code Generator screen, 68, 69f
coding trade-off (simpler code vs. compact, complex
  code), 165
comments, 96
commercial proprietary software, 29–30
COMPARATOR module, 191, 191f
compare capture registers (CCRx), 214, 216, 217f
complete 32-bit STM address range, 103, 104f
Configuration View icons, 64f
Configuration View tab, 63–64
Configure Flash Tools, 82
context switch, 132
core memory address ranges, 105t
core memory addresses, 103–105
Cortex M-4 processor, 4, 5f

Cortex Microcontroller Interface Standard (CMSIS).
  See ARM Cortex Microcontroller Interface
  Standard (CMSIS)
CounterMode, 148
counters, 16–19. *See also* timers
cross-compiler, 30, 34
CTS. *See* clear-to-send (CTS) line
CubeMX, 37–40, 53–86
  ARM CMSIS. *See* ARM Cortex Microcontroller
    Interface Standard (CMSIS)
  Board Selector dialog screen shot, 55, 57f
  Clock Configuration tab, 62–63
  Configuration View tab, 63–64
  features, 53–54
  Help document, 54–55, 56f
  IP tree pane, 61–62
  Load Project, 54, 55f
  MCU alternative functions, 60–61
  MCU-centric utility, 53
  New Project. *See* New Project (CubeMX)
  NUCLEO-F302R8 board selection, 55, 57f
  opening screen, 37f, 54f
  PC13 alternative pin mappings, 60f
  Pinout tab, 58–60
  Power Consumption Calculator View tab, 65
  purpose, 86
  RCC pin configurations, 63

DAC. *See* digital-to-analog converter (DAC)
DAC clock source. *See* digital-to-analog converter
  (DAC) clock source
DAC_ALIGN_12B_R, 270
DAC_ChannelConfTypeDef, 262
DAC_HandleTypeDef, 261
DAC_TRIGGER_NONE, 262
DAC_TRIGGER_SOFTWARE, 262
DAC_TRIGGER_Tx_TRGO, 262
DAC_TypeDef *Instance, 261
daemon, 44
DAP. *See* debug access port (DAP)
data transfer, 241
data transfer concepts, 242–245
data transfer process, 241
data transfer protocol, 168–169, 188
dead time generator, 146
dead time insertions, 146
debug access port (DAP), 79
debugging, 34
debugging tool. *See* OpenOCD15
decrementing counter, 145
define statements, 99, 122
*Definitive Guide to ARM Cortex-M3 and Cortex-M4
  Processors* (Yiu), 4
Device Database menu selection, 71, 71f
Device not found dialog box, 71, 71f
Dhrystone million instructions per second
  (DMIPS), 65
digital-to-analog converter (DAC), 241, 259–278
  analog voltage outputs, 260, 260t, 261f
  buffer output amplifier, 260

HAL framework, 261–262
noise, 260
R-2R resistor ladder network, 259, 259f
sine wave signals, 269–278
triangular waveform, 263–269
digital-to-analog converter (DAC) clock source, 146
direct memory access (DMA), 241–258
  analogy (bypass highway), 243
  bus controller, 241, 244
  bus matrix, 244, 244f, 245
  data transfer concepts, 242–245
  DMA block diagram, 242f
  DMA channels, 245, 247f
  DMA controller details, 245–246
  final steps in setting up DMA transfer, 250–251
  HAL framework, 246–250
  memory port, 245
  peripheral port, 245
  priority assignment hardware unit, 245
  Round Robin algorithm, 244
  slave port, 245
  software triggering, 245
  USART2 sending character string to memory,
    251–258
direct X1 probe settings, 127
DisableInterrupts ( ) 131
disclaimer, 72, 73f
DMA. *See* direct memory access (DMA)
DMA block diagram, 242f
DMA channel block diagram, 247f
DMA channels, 245, 247f
DMA_Channel_TypeDef *Instance, 246
DMA_CIRCULAR, 250
DMA_HandleTypeDef, 246
DMA_HandleTypeDef *DMA_Handle, 192, 193
DMA_HandleTypeDef *DMA_Handle1 261, 262
DMA_HandleTypeDef *DMA_Handle2 261, 262
DMA_InitTypeDef, 248
DMA_InitTypeDef Init, 246, 248
DMA_MDATAALIGN_BYTE, 249
DMA_MDATAALIGN_HALFWORD, 249
DMA_MDATAALIGN_WORD, 249
DMA_MEMORY_TO_MEMORY, 249
DMA_MEMORY_TO_PERIPH, 249
DMA_NORMAL, 250
DMA_PDATAALIGN_BYTE, 249
DMA_PDATAALIGN_HALFWORD, 249
DMA_PDATAALIGN_WORD, 249
DMA_PERIPH_TO_MEMORY, 249
DMA_PINC_DISABLE, 249
DMA_PINC_ENABLE, 249
DMA_PRIORITY_HIGH, 250
DMA_PRIORITY_LOW, 250
DMA_PRIORITY_MEDIUM, 250
DMA_PRIORITY_VERY_HIGH, 250
DMIPS. *See* Dhrystone million instructions per second
  (DMIPS)
Drivers/CMSIS, 70f, 80
Drivers/STM32F3xx_HAL_Driver, 70f, 80
duty cycle, 214

EEPROM, 31
EnableInterrupts ( ), 131
en.stsw-link004, 41
en.stsw-link009, 41
error dialog, 90–91, 92f
E5V external power supply, 25, 25t, 26
exceptions, 129. *See also* interrupt
external interrupt, 134–135
external interrupt controller (EXTI), 16, 134
external power supply output, 26
EXTI. *See* external interrupt controller (EXTI)
EXTI block diagram, 135f
EXTI line [15:10] interrupt, 143
EXTI15_10_IRQ, 135

first microcontroller (TMS1000), 30
flash memory, 105t
floating, 111
floating point unit (FPU), 132
forever loop, 78, 97, 98
FPU. *See* floating point unit (FPU)
full-duplex communications, 168, 169f

GDB. *See* GNU Project Debugger (GDB)
general purpose (GP) timers, 17–18, 17t,
    146, 147t
general purpose input-output (GPIO), 16
    CubeMX, 58–60
    HAL. *See* GPIO pins and hardware abstract layer
        (HAL)
    interrupt, 134–135, 136f
general-purpose timer PWM signal generation,
    214–216
generic toolchain workflow, 33f, 34
GNU Project Debugger (GDB), 44–45
gnuarmeclipse-openocd-win64-0.10.0-201701241841-
    setup.exe, 43, 44f
GP timer. *See* general purpose (GP) timers
GPIO. *See* general purpose input-output (GPIO)
GPIO clock speed, 124–127
GPIO_InitStruct.Mode = GPIO_MODE_
    IT-RISING, 141
GPIO_Instruct, 117
GPIO_Instruct C data structures, 125–126
GPIO_Instruct to GPIO Register Equivalence, 117f
GPIO_Instruct.Mode, 117f
GPIO_Instruct.Pin, 117f
GPIO_Instruct.Pull, 117f
GPIO_Instruct.Speed, 117f
GPIO pin hardware, 109–111
GPIO pin signal waveforms, 126, 126f
GPIO pins and hardware abstract layer (HAL),
    101–128
    core memory addresses, 103–105
    HAL_GPIO. *See* HAL_GPIO module
    memory-mapped peripherals, 103
    peripheral memory addresses, 105–108
GPIO port mode register, 111f
GPIO port pin block diagram/schematic, 109f
GPIO registers, 111, 112f, 113f

GPIO signal skew, 124–125, 128
GPIOA_MODER, 112f
GPIOA_OSPEEDER, 112f
GPIOA_PUPDR, 113f
GPIOB_MODER, 112f
GPIOB_OSPEEDER, 112f
GPIOB_PUPDR, 113f
GPIOx_AFRH, 113f
GPIOx_AFRL, 113f
GPIOx_BSRR, 113f
GPIOx_IDR, 113f
GPIOx_LCKR, 113f
GPIOx_MODER, 110, 112f, 117f
GPIOx_ODR, 113f
GPIOx_OSPEEDER, 112f, 117f
GPIOx_OTYPER, 112f, 117f
GPIOx_PUPDR, 113f, 117f
group, 198

HAL. *See* GPIO pins and hardware abstract layer
        (HAL); HAL framework
HAL framework
    analog-to-digital conversion, 192–197
    digital-to-analog converter (DAC),
        261–262
    direct memory access (DMA), 246–250
    pulse-width modulation (PWM), 216–220
HAL_ADC_Init ( ), 201
HAL_ADC_PollForConversion (&hadc1,
    100), 206
HAL_ADC_Start ( ), 201
HAL_ADC_Start (&hadc1), 206
HAL_ADC_Stop ( ), 201
HAL_DAC_SetValue, 268
HAL_Delay 200, 180, 181
HAL_Delay 500, 98, 99
HAL_Delay 1000, 206, 207
HAL_DMA_IRQHandler ( ), 248
HAL_GPIO module, 107–124
    clock speed demonstration, 124–127
    GPIO pin hardware, 109–111
    GPIO port mode register, 111f
    GPIO registers, 111, 112f, 113f
    main.c code listing, 117–121
    modification (enabling multiple LED outputs),
        122–123
    push button test demonstration, 123–124, 128
    simple LED test demonstration, 111–122, 128
HAL_GPIO_TogglePin (LD2_GPIO_Port,
    LD2_Pin), 98, 99
HAL_Init ( ), 78
HAL_LockTypeDef Lock, 192, 193, 246, 248,
    261, 262
HAL_NVIC_EnableIRQ (EXTI15_10_IRQn),
    141
HAL_NVIC_EnableIRQ (TIM2_IRQn), 156
HAL_NVIC_SetPriority (ETXTI15_10_IRQn,
    0, 0), 141
HAL_NVIC_SetPriority (TIM2_IRQn, 0, 0),
    156

HAL_StatusTypeDef_HAL_ADC_
   PollForConversion (ADC_Handle_
   TypeDef* hadc, uint32_t Timeout),
   201
HAL_StatusTypeDef_HAL_DAC_Start_DMA
   (DAC_HandleTypeDef *hdac, uint32_t
   Channel, uint32_t* pData, uint32_t
   Length, uint32_t Alignment), 269
HAL_StatusTypeDef_HAL_DMA_Start
   (DMA_HandleTypeDef *hdma, uint32_t
   SrcAddress, uint32_t Dst Address,
   uint32_t DataLength), 250
halt command, 51, 51f
HAL_TIM_Base_Start_IT (&htim2), 156
HAL_UART_Transmit (&huart2, (uint8_t*)
   msg, 10, 100), 180, 181
HAL_UART_Transmit (&huart2, (uint8_t*)
   msg, 20, 100), 206, 207
hardware, 1–28. See also microcontroller (MCU);
   Nucleo-64 board options
hardware abstract layer. See GPIO pins and hardware
   abstract layer (HAL)
hardware timers. See timers
Harvard architecture, 5
header file, 97
Hello World program. See STM project development
hex code download, 82–86
high-resolution timers, 146
high-speed external clock (HSE), 23–24, 62–63, 270
historical overview, 30–31
Hitec model HS-311 analog servo, 236f, 237
hobby-grade servo motor, 235–239
HRTIM1, 146
HSE. See high-speed external clock (HSE)
human eye brightness perception, 229, 230f
HwFlowCtl, 172

I2C, 19–20
I2C block diagram, 20f
I2C signal lines, 20f
I2S, 21
I2S2, 21
I2S3, 21
I/O pin, 109, 109f
IDE. See integrated development environment (IDE)
IDE download Web page, 36f
if/else statement, 99, 100
IN0, 198, 199f
IN1, 198, 199f
#include "main.h", 92, 96–97
infinite loop, 97. See also forever loop
Init, 170
Instance, 170
int, 97
integrated development environment (IDE), 32
integrated peripheral (IP) tree pane, 61–62
Intel 805x family of MCUs, 31
Intel 8048, 30
Intel 8052, 31, 32t
Intel 8751, 31, 31f

inter-integrated circuit. See I2C
inter-integrated sound. See I2S
interrupt, 129–144
   acknowledgment instruction, 133
   arming/disarming, 131
   asynchronous event, 129
   background thread/foreground thread, 132
   BASEPRI register, 131–132
   context switch, 132
   defined, 14, 129
   demonstration/test run, 135–144
   enabling/disabling, 131
   external, 134–135
   EXTI line [15:10], 143
   EXTI15_10_IRQ, 135
   GPIO pins, external interrupt controller, and NVIC
      IRQs, 134–135, 136f
   main.c file listing, 137–141
   maskable, 131
   MX_GPIO_Init method, 138, 141
   NVIC, 130–131
   overflow, 145
   pending interrupt request, 133
   polling, 133–134
   preconditions, 132
   priority level, 15, 131
   processing sequence, 132, 133f
   RCC global, 142–143
   stm32f3xx_it.c file listing, 142
   System tick timer, 142
   SysTick, 131, 133
   USART2 global, 182
   users, 129
   vector table, 133
   what happens when interrupt is recognized, 132,
      133f
interrupt channel, 130, 131
interrupt-driven blink LED timer, 155–157
interrupt flow diagram, 14f
interrupt handler, 129
interrupt processing sequence, 132, 133f
interrupt request (IRQ), 15, 130, 130f
interrupt service routine (ISR), 14, 132–134, 143
_IO HAL_DAC_StateTypeDef State, 261, 262
_IO HAL_DMA_StateTypeDef State, 246, 248
_IO uint32_t, 192, 193
_IO uint32_t ErrorCode, 193, 246, 248,
   261, 262
_IO uint32_t State, 193
I/O pin, 109, 109f
Ionescu, Liviu, 43
IOREF pin, 11
IP tree pane. See integrated peripheral (IP) tree pane
IRQ. See interrupt request (IRQ)
ISR. See interrupt service routine (ISR)

JP1, 25
JP1 configuration table, 25t
JP5 jumper positions, 26t
JP6, 23

Keil IDE, 30, 32, 35–36, 52, 90, 92f
Keil IDE build icon, 81, 81f
Keil IDE v4.74 icon, 36f
Keil registration form, 35f

LD1, 22
LD2, 22
LD3, 22
leading edge delay, 124
leading edge rise time, 127
LED
  color/luminescence of tricolor LED, 231–235
  interrupt-driven blink LED timer, 155–157
  LED states, 22
  linear dimming property, 231
  luminescence of normal LED, 228–231
  multi-rate interrupt-driven blink LED timer,
    157–164, 166
  non-interrupt blink LED timer, 149–155
  Nucleo-64 board options, 22
  polled blink LED timer, 149–155
light-emitting diode. See LED
loader, 34
Lock, 170
low-ass analog filter, 260
low clock speed, 127
low power timers, 147
low-speed external clock (LSE), 24, 62–63
LR register, 132
LSE. See low-speed external clock (LSE)

maskable interrupt, 131
master clock, 145
master clock output (MCO), 23
Mastering STM32 (Noviello), 53
maximum ambient temperature (TAMAX), 65
MCO. See master clock output (MCO)
MCU. See microcontroller (MCU)
MCU pinout diagram, 40, 40f
MCU toolchain, 32–52
  debugging tool. See OpenOCD15
  integrated development environment (IDE), 32
  Keil IDE, 35–36
  ST-LINK, 40–43
  STMCubeMX, 37–40
  toolchain workflow, 33f, 34
mdb command, 52, 52f
mdh command, 52, 52f
MDK-ARM subdirectory, 82
MDK-ARM V4, 88
MDK474.exe, 36
mdw command, 52, 52f
memory, 5–6
memory management unit (MMU), 103
memory-mapped peripherals, 103
memory mapping, 103
memory port, 245
memory protection unit (MPU), 105
memory-to-memory transfer, 244
memory-to-peripheral transfer, 244

microcomputer, 1
microcontroller (MCU)
  analog-to-digital converter (ADC), 16
  bit serial ports, 19–21
  components, 3–21
  defined, 1
  external interrupt/event controller (EXTI), 16
  general purpose input-outputs (GPIOs), 16
  historical overview, 30–31
  I2C serial protocol, 19–20
  inter-integrated sound (I2S) interface, 21
  memory, 5–6
  microprocessor, compared, 1
  nested vector interrupt controller (NVIC), 14–16
  peripherals, 6–13
  processor, 3–5
  SPI serial protocol, 20–21
  timers and counters, 16–19
  USART serial protocol, 19
microprocessor, 1
MiddleWares, 58
MMU. See memory management unit (MMU)
Mode, 172
MODER, 111
modulus operator, 165
Morpho connectors, 13, 114f
MPU. See memory protection unit (MPU)
multi-channel oscilloscope, 125
multi-channel scan/ continuous conversion, 198
multi-channel scan/single conversion, 197–198, 198f
multi-rate interrupt-driven blink LED timer, 157–164,
    166
multiple brace pairs, 97
MX_GPIO_Init ( )
  CubeMX, 78
  interrupt, 138, 141
  STM project development, 93, 97
  timers, 155

N-BIT REGISTER, 190–192
nested vector interrupt controller (NVIC), 14–16, 129,
    130–131
nesting, 97
New C/C++ Project dialog box, 88f
New Project (CubeMX), 55, 66–78
  Application/MDK-ARM, 70f
  Application/User, 70f
  Code Generation dialog box, 68, 69f
  Code Generator screen, 68, 69f
  Device Database menu selection, 71, 71f
  Device not found dialog box, 71, 71f
  Drivers/CMSIS, 70f
  Drivers/STM32F3xx_HAL_Driver, 70f
  HAL_Init ( ), 78
  main.c code listing, 72–78
  MX_GPIO_Init ( ), 78
  Open Project button, 68
  opening screen, 66f
  Project dropdown menu, 66, 67f
  Project Settings screen, 67–68, 67f

Project Settings submenu, 68
STM disclaimer, 72, 73f
STM32F302RB selection, 72, 72f
STM32Nucleo_F302 project directories, 68, 70f
SystemClock_Config ( ), 78
while (1) statement, 78
New Project opening screen, 66f
NMI. *See* non-maskable interrupt (NMI)
noise, 260
non-interrupt blink LED timer, 149–155
non-maskable interrupt (NMI), 130, 130f
Noviello, Carmine, 53
NRST, 11, 58f, 59
Nucleo board user manual, 113
Nucleo boards, 2–3
Nucleo-32 board, 3f
Nucleo-64 block diagram, 7f
Nucleo-64 board, 3, 3f
Nucleo-64 board layout, 6–13
Nucleo-64 board options, 22–27
    JP6, 23
    LEDs, 22
    OSC 32-kHz clock supply, 24
    OSC clock, 23–24
    power supply and power selection, 24–27
    push buttons, 22–23
Nucleo-64 NUCLEO-F302R8 board, 39, 39f
Nucleo-64 pinout diagram, 10f
Nucleo-144 board, 3f
NUCLEO-F302R8 board, 39, 39f
NUCLEO-F302R8 board selection, 55, 57f
NUCLEO-F411RE development board, 58
NVIC. *See* nested vector interrupt controller (NVIC)
NVIC BASEPRI register, 131–132
NVIC block diagram, 130f

Open On-Chip Debugger, 43. *See also* OpenOCD15
Open Project button, 68
Open-source vs. commercial proprietary software, 29–30
opening and closing braces { }, 96, 97
OpenOCD15, 43–52
    accessing the program, 44
    finish screen, 45f
    GNU Project Debugger (GDB), 44–45
    gnuarmeclipse-openocd-win64-0.10.0-
        201701241841-setup.exe, 43, 44f
    help request, 49, 49f
    memory content display commands, 51–52, 52f
    opening screen, 45f
    reg command after a halt, 51, 51f
    reg command after a reset, 50f, 51
    service or background program, 44
    starting OpenOCD service, 47f
    Telnet link, 46–48
    user guide, 49–50
    versions, 43, 44f
    Zadig utility, 47–48
OpenOCD15 downloads, 44f
OpenOCD15 finish screen, 45f

OpenOCD15 opening screen, 45f
OpenOCD15 user guide, 49–50
OSC 32-kHz clock supply, 24
OSC clock, 23–24
output capture, 17, 146
overclocking, 63
overflow interrupt, 145
OverSampling, 172

PA2, 19, 58f, 59, 174
PA3, 19, 58f, 59, 174
PA6, 122
PA10, 115, 117, 122
Parity, 172
PB13, 58f, 59
PC. *See* program counter (PC)
PC13, 58f, 59
PC13 alternative pin mappings, 60f
PC register, 132
PCC. *See* power consumption calculator (PCC)
PDSC, 79
pending interrupt request, 133
peripheral memory addresses, 105–108
peripheral port, 245
peripheral-to-memory transfer, 244
peripheral-to-peripheral transfer, 244
peripherals, 6–13
phase-locked loop (PLL), 145
Picoscope USB oscilloscope, 125
pin clock speeds, 125–126
pinout diagram, 40, 40f
Pinout tab, 58–60
PLL. *See* phase-locked loop (PLL)
polled blink LED timer, 149–155
polling
    analog-to-digital conversion, 201–212
    interrupts, 133–134
Port, 173
Ports (COM & LPT), 173, 174
power consumption calculator (PCC), 65f
Power Consumption Calculator View tab, 65
power-on sequence procedure, 27
power supply and power selection, 24–27
preemption and interrupt nesting, 15f
prescaler, 145, 147–148
prioritized interrupt, 15, 131
priority assignment hardware unit, 245
processor, 3–5
Program and Verify dialog box, 85f
program counter (PC), 14
project build, 100. *See also* STM project development
Project dropdown menu, 66, 67f, 88, 89f
Project pinout screen, 88, 89f
Project Settings/Code Generator dialog box, 89–90, 91f
Project Settings dialog box, 88–89, 90f
Project Settings screen, 67–68, 67f
Project Settings submenu, 68
pRxBuffPtr, 170
pTxBuffPtr, 170

pulse-width modulation (PWM), 213–240
  50% duty cycle waveform, 227–228, 229t
  CIE1931 lightness formula, 231
  color and luminescence of tricolor LED, 231–235
  duty cycle, 214
  enabling PWM function, 220–222
  frequency, 213–214
  general-purpose timer PWM signal generation,
    214–216
  HAL framework, 216–220
  hobby-grade servo motor, 235–239
  human eye brightness perception, 229, 230f
  luminescence of normal LED, 228–231
  purposes, 213
  PWM demonstration software, 222–227
  timer hardware architecture, 216, 217f
  typical PWM signal, 214f
push button test demonstration, 123–124, 128
push buttons, 22–23
push pins, 61
PWM. *See* pulse-width modulation (PWM)

R-2R resistor ladder network, 259, 259f
rank, 199, 199f
Rath, Dominic, 43
raw = (double) HAL_ADC_GetValue
  (&hadc1), 206
raw = raw * 0.452, 206
RCC global interrupt, 142–143
RCC pin configurations, 63
real-time operating system (RTOS), 18, 79, 105
Realterm Port screen, 173f
Realterm terminal program, 172–175
reference manual, 101
reg command, 50f, 51, 51f
Reinstall WinUSB driver dialog box, 48f
RepetitionCounter, 148
request/acknowledge, 169
request-to-send (RTS) line, 169
reset command, 50f, 51
Round Robin algorithm, 244
RTOS. *See* real-time operating system (RTOS)
RTS. *See* request-to-send (RTS) line
RxXferCount, 170

sample/hold circuit, 190
SAR. *See* successive-approximation register (SAR)
SAR LOGIC module, 191, 191f
SB13, 19
SB14, 19
SB62, 19, 175
SB63, 19, 175
serial peripheral interface. *See* SPI
servo motor, 235–239
signal waveforms, 126, 126f
simple 16-bit timer, 17t, 18, 146, 147t
simpler code vs. compact, complex code, 165
sine wave signals, 269–278
single channel/continuous conversion, 198
single channel/single conversion, 197, 197f

slave port, 245
software. *See* STM MCU software
source code, 34
SPI, 20–21
SPI block diagram, 21f
SPI2 signal lines, 21t
sprintf (msg, "%frn," value), 180, 206, 207
SRAM, 105t
SSI. *See* synchronous serial interface (SSI)
ST-LINK, 40–43, 82–86
ST-LINK firmware version display, 43f
ST-LINK/V2 opening screen, 41f
static void MX_ADC1_Init (void), 204
static void MX_TIM2_Init (void), 220, 221
static void MX_USART2_UART_Init (void),
  175, 176
stlink_winusb_install, 41
STM. *See* STMicroelectronics (STM)
STM core memory address ranges, 105t
STM disclaimer, 72, 73f
STM GPIO port pin block diagram/schematic, 109f
STM MCU databases, 2
STM MCU interrupt, 144. *See also* interrupt
STM MCU software, 29–52
  bare metal development, 30
  open-source vs. commercial proprietary software,
    29–30
  toolchain. *See* MCU toolchain
STM Nucleo board user manual, 113
STM Nucleo boards, 2–3
STM Nucleo-64 microcontroller, 27. *See also*
  microcontroller (MCU)
STM project development, 87–100
  complex modification of main.c file, 99–100
  error dialog, 90–91, 92f
  explanation of C/C++ code, 96–97
  Keil IDE, 90, 92f
  main.c file, 91–97
  MDK-ARM V4, 88
  New C/C++ Project dialog box, 88f
  Project dropdown menu, 88, 89f
  Project pinout screen, 88, 89f
  Project Settings/Code Generator dialog box, 89–90,
    91f
  Project Settings dialog box, 88–89, 90f
  simple modification of main.c file, 98–99
  Successful Code Generation dialog box, 90, 91f
STM timer peripherals, 145–147. *See also* timers
STM32-ST-LINK Utility V4.0.0 setup application, 41
STM32 toolchain, 33. *See also* MCU toolchain
STM32CubeMX, 37–40, 53–86. *See also* CubeMX
STM32CubeMX opening splash screen, 37f, 54f
STM32F302R8 block diagram, 101, 102f
STM32F302R8 datasheet, 101
STM32F302R8 MCU, 6, 27, 32t
STM32F302R8 MCU pinout diagram, 40, 40f
STM32F302R8 reference manual, 101
STM32F302RB selection, 72, 72f
STM32F302R8Tx MCU, 58f
stm32f3xx_it.c file listing, 142

STMicroelectronics (STM), 29
STM32Nucleo_F302 project directories, 68, 70f
STM32Toolchain directory, 37
st_nucleo_f3.cfg, 46, 46t
StopBits, 171
STSW-LINK004, 41
STSW-LINK009, 40
Successful Code Generation dialog box, 90, 91f
successive-approximation register (SAR), 190, 191f
SVD. See System View Description (SVD)
synchronous serial data link, 20
synchronous serial interface (SSI), 20
synthesizing a sine wave, 270
system memory, 105t
System View Description (SVD), 79
SystemClock_Config ( ), 78, 93, 97
SysTick interrupt, 131, 133
SysTick timer, 18–19

tail-chaining, 15f
TAMAX. See maximum ambient temperature
    (TAMAX)
Telnet link, 46–48
Telnet scripts, 46t
terminal program, 172–175
THD. See total harmonic distortion (THD)
TIM1. See advanced control timer (TIM1)
TIM2, 17, 17t, 147t
TIM2 clock source selection, 149f
TIM6, 17t, 18, 147t
TIM15, 17, 17t, 147t
TIM16, 17, 17t, 147t
TIM17, 17, 17t, 147t
TIM_CLOCKDIVISION_DIV, 148
TIM_CLOCKDIVISION_DIV2, 148
TIM_CLOCKDIVISION_DIV4, 148
TIM_COUNTERMODE_CENTERALIGNED1, 148
TIM_COUNTERMODE_CENTERALIGNED2, 148
TIM_COUNTERMODE_CENTERALIGNED3, 148
TIM_COUNTERMODE_DOWN, 148
TIM_COUNTERMODE_UP, 148
timer counter (CNT), 216, 217f
timer hardware architecture, 216, 217f
timer update register, 148
timers, 16–19, 145–166
    categories, 146–147
    configuration, 147–149
    counting up/counting down, 145
    decrementing counter, 145
    definition, 145
    hardware architecture, 216, 217f
    interrupt-driven blink LED timer, 155–157
    manufacturer datasheets, 147
    master clock, 145
    multi-rate interrupt-driven blink LED timer,
        157–164, 166
    non-interrupt blink LED timer, 149–155
    overflow interrupt, 145
    parameters and settings, 147–148
    polled blink LED timer, 149–155

STM32F302R8 MCU timers, 17t, 147t
timer update register, 148
trade-off (simpler code vs. compact, complex
    code), 165
update event calculation, 148–149
uses, 146
TIM_HandleTypeDef, 147, 155
TIM_MasterConfigTypeDef, 216
TIM_MASTERSLAVEMODE_DISABLE, 216
TIM_MASTERSLAVEMODE_ENABLE, 216
TIM_OCIDLESTATE_RESET, 219
TIM_OCIDLESTATE_SET, 219
TIM_OC_InitTypeDef, 218
TIM_OCMODE_ACTIVE, 218
TIM_OCMODE_FORCED_ACTIVE, 218
TIM_OCMODE_FORCED_INACTIVE, 218
TIM_OCMODE_INACTIVE, 218
TIM_OCMODE_PWM1-PWM Mode 1, 218
TIM_OCMODE_PWM2-PWM Mode 2, 218
TIM_OCMODE_TIMING, 218
TIM_OCMODE_TOGGLE, 218
TIM_OCNFAST_DISABLE, 219
TIM_OCNFAST_ENABLE, 219
TIM_OCNIDLESTATE_RESET, 220
TIM_OCNIDLESTATE_SET, 220
TIM_OCNPOLARITY_HIGH, 219
TIM_OCNPOLARITY_LOW, 219
TIM_OCPOLARITY_HIGH, 219
TIM_OCPOLARITY_LOW, 219
TMP36 temperature sensor, 189, 190f, 202–205
TMS1000, 30
toggle method, 165
toolchain, 32. See also MCU toolchain
toolchain workflow, 33f, 34
total harmonic distortion (THD), 277
track/hold circuit, 190
trade-off (simpler code vs. compact, complex code),
    165
trailing edge delay, 124
tree pane IP icons, 61f
triangular waveform, 263–269
trivial bug, 34
TxXferCount, 170

UART, 167
UART block diagram, 167f
UART vs. USART, 167, 168
UART_HandleTypeDef, 170
UART_HWCONTROL_CTS, 172
UART_HWCONTROL_NONE, 172
UART_HWCONTROL_RTS, 172
UART_HWCONTROL_RTS_CTS, 172
UART_InitTypeDef, 170, 171
UART_MODE_RX, 172
UART_MODE_TX, 172
UART_MODE_TX_RX, 172
UART_OVERSAMPLING_8, 172
UART_OVERSAMPLING_16, 172
UART_PARITY_EVEN, 172
UART_PARITY_NONE, 172

UART_PARITY_ODD, 172
UART_STOPBITS_1, 171
UART_STOPBITS_2, 171
UART_WORDLENGTH-8B, 171
UART_WORDLENGTH-9B, 171
uint32_t Channel, 199
uint32_t ClockPrescaler, 193, 194
uint32_t ContinuousConvMode, 193, 195
uint32_t DAC_OutputBuffer, 262
uint32_t DAC_Trigger, 262
uint32_t DataAlign, 193, 194
uint32_t Direction, 249
uint32_t DiscontinuousConvMode, 193, 195
uint32_t DMAContinuousRequests, 194, 196
uint32_t EOCSelection, 193, 195
uint32_t ExternalTrigConv, 194, 195-196
uint32_t ExternalTrigConvEdge, 194, 196
uint32_t HAL_ADC_GetValue (ADC_
    HandleTypeDef* hadc), 201
uint32_t MasterOutputTrigger, 216
uint32_t MasterSlaveMode, 216
uint32_t MemDataAlignment, 249
uint32_t MemInc, 249
uint32_t Mode, 249, 250
uint32_t NbrOfConversion, 193, 195
uint32_t OCFastMode, 218, 219
uint32_t OCIdleState, 218, 219
uint32_t OCMode, 218
uint32_t OCNIdleState, 218, 220
uint32_t OCNPolarity, 218, 219
uint32_t OCPolarity, 218, 219
uint32_t Offset, 199, 200
uint32_t PeriphDataAlignment, 249
uint32_t PeriphInc, 249
uint32_t Priority, 249, 250
uint32_t Pulse, 218, 219
uint32_t Rank, 199-200
uint32_t Resolution, 193, 194
uint32_t SamplingTime, 199, 200
uint32_t ScanConvMode, 193, 194
uint32_t* Alignment, 269, 270
uint32_t* Length, 269, 270
uint32_t* pData, 269, 270
universal synchronous/asynchronous receiver/
    transmitter. See USART

universal synchronous receiver transmitter.
    See UART
update event calculation, 148-149
USART, 19, 167, 168, 168f
USART block diagram, 168f
USART configuration, 170-172
USART receive demonstration program, 182-188
USART_RX, 174, 174f
USART transmit demonstration program, 176-181
USART_TX, 174, 174f
USART1, 169
USART2, 19, 169
USART2 global interrupt, 182
USART2 sending character string to memory,
    251-258
USART3, 19
USB cable, 42, 42f
USB oscilloscope, 125, 227
USB ST-LINK driver software, 40
U5V, 13
UV EPROM, 31
uVision, 36

value = (raw - 500.0)/10, 206
value = (9*value) /5 + 32, 206
VBAT input, 13
vector interrupt list, 15
very fast clock speed, 127
VIN, 11
VIN external power supply, 25, 25t, 26
void *Parent, 246, 248
void (* XfercpltCallback) (struct
    _DMA_HandleTypeDef *hdma), 246, 248
void (* XferErrorCallback) (struct
    _DMA_HandleTypeDef *hdma), 246, 248
void EXTI15_10_IRQHandler (void),
    141-142
von Neumann architecture, 5

watchdog timers, 18
while (1) statement, 78, 93, 97, 98
Windows terminal program, 172-175
Wordlength, 171

Zadig utility, 47-48